Springer Topics in Signal Processing

Volume 10

For further volumes:
http://www.springer.com/series/8109

Amar Mitiche · J. K. Aggarwal

Computer Vision Analysis of Image Motion by Variational Methods

Springer

Amar Mitiche
INRS-Energie, Matériaux et Télécommunications
Institut National de la Recherche Scientifique
Montreal, QC
Canada

J. K. Aggarwal
Department of Electrical and Computer Engineering
The University of Texas
Austin, TX
USA

ISSN 1866-2609 ISSN 1866-2617 (electronic)
ISBN 978-3-319-00710-6 ISBN 978-3-319-00711-3 (eBook)
DOI 10.1007/978-3-319-00711-3
Springer Cham Heidelberg New York Dordrecht London

Library of Congress Control Number: 2013939338

Printed on acid-free paper

Springer is part of Springer Science+Business Media (www.springer.com)

Contents

Chapter 1
Image Motion Processing in Visual Function

Retinal motion comes about whenever we move or look at moving objects. Small involuntary retinal movements take place even when we fixate on a stationary target. Processing of this ever-present image motion plays several fundamental functional roles in human vision. In machine vision as well, image motion processing by computer vision algorithms has in many useful applications several essential functions reminiscent of the processing by the human visual system. As the following discussion sets to point out, computer vision modelling of motion has addressed problems similar to some that have arisen in human vision research, including those concerning the earliest fundamental questions and explanations put forth by Helmholtz and by Gibson about human motion perception. However, computer vision motion models have evolved independently of human perception concerns and specificities, much like the camera has evolved independently of the understanding of the human eye biology and function [1].

1.1 Image Motion in Visual Function

The most obvious role of image motion processing by the human visual system is to perceive the *motion of real objects*. The scope and quality of this perception varies widely according to the visual task performed, ranging from detection where moving versus static labelling of objects in the visual field is sufficient, to event interpretation where a characterization of motion by more detailed evaluation or attributes is required.

Less evident a role is the perception of *depth*. Computational and experimental investigations have revealed the link between the image motion and the variables of depth and three-dimensional (3D) motion. To emphasize this role of image motion, Nakayama and Loomis [2] named *kineopsis*, by analogy to stereopsis, the process of recovering depth and 3D motion from image motion.

A. Mitiche and J. K. Aggarwal, *Computer Vision Analysis of Image Motion by Variational Methods*, Springer Topics in Signal Processing 10, DOI: 10.1007/978-3-319-00711-3_1,

Kineopsis: The role of motion in the perception of depth, and structure thereof, has been known for a long time. In the words of Helmholtz for instance ([3], pp. 297), over a hundred years ago in his *Handbook of Physiological Optics*, 1910:

> "If anybody with two good eyes will close one of them and look at unfamiliar objects of irregular form, he will be apt to get a wrong, or at any rate an unreliable, idea of their shape. But the instant he moves about, he will begin to have the correct apperceptions."

He adds the following explanation as to the origin of this perception of environmental structure, or apperception as he called it:

> "In the variations of the retinal image, which are the results of movements, the only way an apperception of differences of distance is obtained is by comparing the instantaneous image with the previous images in the eye that are retained in memory."

This is the first recorded enunciation of *structure-from-motion*, tying the perception of structure to image brightness variations. By distinguishing geometry from photometry, Gibson elaborated on this Helmholtz view of structure-from-motion and stated in his book *The Perception of the Visual World*, 1950, that image motion was the actual stimulus for the perception of structure, rather than image variations as Helmholtz conjectured. He was quite explicit about it when he wrote ([4], pp.119):

> "When it is so considered, as a projection of the terrain or as the projection of an array of slanted surfaces, the retinal image is not a picture of objects but a complex of variations. If the relative motion is analyzed out and isolated from the complex of other variations, it proves to be a lawful and regular phenomenon. Defined as a *gradient* of motion, it is potentially a stimulus correlate for an experience of continuous distance on a surface, as we shall see, and one no longer is required to postulate a process of unconscious inference about isolated objects."

By gradient of motion Gibson meant not the spatial or temporal variations of image motion but the image motion field itself, or *optical flow*, stating, when he discussed the example of the motion field on the retina of a flier landing on a runway ([4], pp. 128), that:

> "The gradients of motion are approximately represented by a set of vectors indicating direction and rate at various points. All velocities vanish at the horizon".

The veracity of Gibson's statement that image motion is the stimulus for the perception of structure is not so much surprising when we observe that the perception of the structure of a surface in motion does not change for different texture coverings of this surface. There have been several experiments designed to demonstrate unequivocally this perception of structure-from-motion, first the landmark *kinetic depth effect* experiment of Wallach and O'Connell [5] which used the shadow of a tilted rod projected on a translucent screen which viewers observed from the back. It was also demonstrated by Gibson et al. [6] who used a texture of paint splashed on two lined-up parallel transparent screens the shadows of which were presented to viewers on a frontal translucent screen. The most striking demonstrations are perhaps the experiments of Ullman [7] and of Rogers and Graham [8] with random dot distributions. Random dots constitute stimuli void of any texture or geometric arrangement. Rogers and Graham's demonstration [8] is to some extent a mechanical counterpart of Ullman's experiment with computer-generated random dots on rotating cylinders [7]. Ullman presented viewers with the orthographic projection on a computer screen of about a hundred points on each of two imaginary coaxial

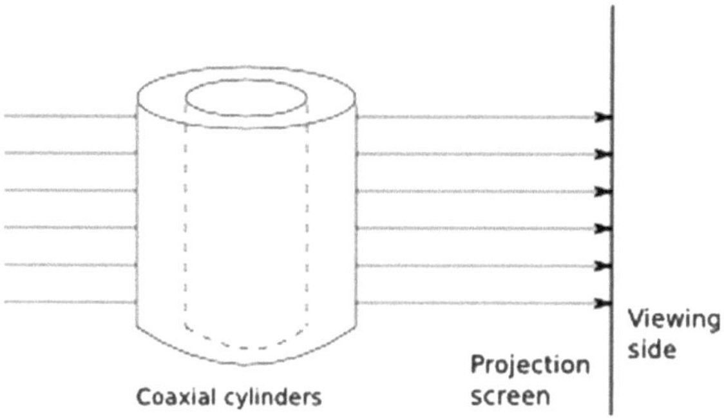

Fig. 1.1 Ullman's rotating cylinders setup simulated by a computer program: Viewers were shown the orthographic projection on a computer screen of a set of about a hundred random points on each of two coaxial cylinders of different radii. The cylinders outline was not included in the display so that they were imaginary to the viewers and, therefore, contributed no clue to the perception. Looking at the random dots image on the screen when the cylinders were not moving afforded no perception of depth. But when the cylinders were made to move, by a computer program, observers reported the vivid perception of two rotating coaxial cylinders and were also able to give a good estimate of the amount of rotation

cylinders of different radii (Fig. 1.1). The cylinders were imaginary in the sense that their outline was not presented in the display so as not to offer viewers a cue to the perception of structure. Looking at the image on the screen of the random dots on static cylinders afforded no perception of depth. But when the cylinders were made to move, by a computer program, observers reported perceiving vividly two rotating coaxial cylinders and could also estimate the amount of rotation.

The view of Helmholtz on the role of image variations in the perception of structure, which we quoted previously, is quite general but otherwise correct, because the link between image variations and image motion is *lawful*, to employ this expression often used by Gibson to mean a relation which can be formally specified by governing laws. Horn and Schunck provided us with such a law [9] in the form of a relation, or equation, deduced from the assumption that the image sensed from a given point of a surface in space remains unchanged when the surface moves. The equation, which we will investigate thoroughly in Chap. 3 and use repeatedly in the other chapters, is called the *optical flow constraint*, or the Horn and Schunck equation:

$$I_x u + I_y v + I_t = 0, \tag{1.1}$$

where I_x, I_y, I_t are the image spatiotemporal derivatives, t being the time and x, y the image spatial coordinates, and $(u, v) = (\frac{dx}{dt}, \frac{dy}{dt})$ is the optical flow vector. The equation is written for every point of the image positional array.

As to the link between image motion and environmental motion and structure, one can get a law, or equation, by drawing a viewing system configuration model, projecting points in three-dimensional space onto the imaging surface, and taking the time derivative of the projected points coordinates. Under a Cartesian reference system and a central projection model of imaging (to be detailed in Chap. 6), one can immediately get equations connecting optical flow to 3D motion and depth:

$$\begin{aligned} u &= f\frac{U - xW}{Z} \\ v &= f\frac{V - yW}{Z}, \end{aligned} \tag{1.2}$$

where X, Y, Z are the 3D coordinates, Z being the depth, and $U = \frac{dX}{dt}, V = \frac{dY}{dt}, W = \frac{dZ}{dt}$ are the corresponding coordinates of the 3D motion vector, called the *scene flow* vector, and f is a constant representing the focal length of imaging. The equation applies to every point on the visible environmental surfaces. The viewing system configuration model, and Eq. (1.2), as well as other equations which follow from it, will be the subject of Chap. 6. We will nonetheless mention now the special but prevailing case of rigid 3D motion, i.e., compositions of 3D translations and rotations. When we express 3D velocity (U, V, W) as coming from rigid body motion, then we have the Longuet-Higgins and Prazdny model equations [10]:

$$\begin{aligned} u &= -\frac{xy}{f}\omega_1 + \frac{f^2 + x^2}{f}\omega_2 - y\omega_3 + \frac{f\tau_1 - x\tau_3}{Z} \\ v &= -\frac{f^2 + y^2}{f}\omega_1 + \frac{xy}{f}\omega_2 + x\omega_3 + \frac{f\tau_2 - y\tau_3}{Z}, \end{aligned} \tag{1.3}$$

where τ_1, τ_2, τ_3 are the coordinates of the translational component of the rigid motion and $\omega_1, \omega_2, \omega_3$ those of the rotational component.

Equations 1.2 and 1.3 reveal an important basic fact: image motion codes depth and 3D motion simultaneously, legitimizing the definition of kineopsis as the process by which depth and 3D motion are recovered from image motion. However, depth can evidently be eliminated from Eqs. (1.2) and (1.3) to obtain an equation which references 3D motion only. Such computational manipulations will be studied in Chap. 6.

Studies of kineopsis have generally distinguished *ego-motion* in a static environment, which is the motion of an observer, or a viewing system, from *object motion*. From a general point of view, kineopsis applies to ego-motion and, as such, image motion from ego-motion in a static environment instructs the observer about its position and motion relative to the surrounding objects. When compared to object motion, the particularity of ego-motion is that it causes image motion everywhere on the image domain whereas object motion induces it only over the extent of the object image. As a result, it provides global information about the observer motion, such as the direction of heading. In machine vision, camera motion is represented by a rigid motion, the parameters of which are, therefore, also global parameters of image motion.

It is apparent from our discussion of kineopsis that image motion processing by computational systems, mainly investigated in computer vision, and by the human visual system, investigated in fields such as psychology, psychophysics, and neurophysiology, play similar roles, although the tools of investigation and analysis to determine and characterize these roles may be quite different. Other important roles include:

Image segmentation: Motion-based image segmentation is a natural, effortless routine activity of humans which allows them to interpret their environment in terms of moving objects and to distinguish one motion from another. In machine vision as well, computational methods have been devised which can be very effective at partitioning an image into different motion regions. In general, computational schemes implement some form of the Gestalt principle of common fate by assuming that the image velocities of points on a moving object are similar and smoothly varying except at occlusion boundaries where sharp discontinuities occur. Motion-based segmentation generally aims at image domain partitioning into distinctly moving objects in space but its scope can reduce to the simpler but nevertheless important case of **motion detection**, which aims at identifying the foreground of moving objects against the background of the unmoving objects without concern about distinguishing between different motions.

Tracking: Image motion processing can assist tracking, the process of following the image of a moving object as it progresses through the visual field. Oculomotor pursuit in the human visual system is driven by velocity information and can adapt to target behaviour. For instance, constant velocity target motion triggers saccadic eye pursuit and accelerated motion is followed by smooth eye movement. In computer vision, target tracking is an essential component of systems in applications such as visual surveillance and monitoring.

Pattern vision and perception: Motion affects pattern vision and perception. For instance, pattern motion can enhance the vision of low spatial frequencies by the human eye and can degrade the vision of high frequencies. In computer vision, patterns of object motion can map to distinctive patterns of velocity which, therefore, can serve to interpret objects dynamic behaviour.

1.2 Computer Vision Applications

Image motion processing plays an essential role in many computer vision applications. Here following are a few of its current important uses.

Robotics: A major goal of robotics is to give camera-equipped mobile robots the ability to use vision processing to navigate autonomously in their environment [11–13]. Image motion analysis can be used to address problems such as robot positioning and guidance, obstacle avoidance, and tracking of moving objects. Visual servoing [14–16] is also of great interest in robotics. Its purpose is to bring a robot to a position where the sensed image agrees with a desired visual pattern. This movement control, which uses visual feedback, can assist an autonomously moving robot in keeping its course on a target in its visual field, or using an end-effector to manipulate an object.

Human activity analysis [17–19]: The characterization and recognition of patterns in human motion are the focus of scientific investigations in several applications. In biomedical imaging, for instance, gait kinematic measurement patterns can serve knee pathology diagnosis and re-education prescription. Other examples include the study of lip movements or hand gestures for visual communication, and human motion capture for computer animation. There is also a considerable interest in visual monitoring applications such as surveillance [20]. Tasks which visual surveillance systems take up include traffic survey, control of access to secured sites, and monitoring of human activity. These tasks require motion detection and some form of motion estimation and tracking.

Video compression [21]: Video data compression is indispensable in digital video services such as telecommunications, broadcasting, and consumer electronics, because enormous amounts of data are continually produced that must be processed and transmitted with imperceptible delay. Current standards of video transmission apply motion compensation, whereby image motion is used to predict and code the temporal progression of an image.

Video description: Video archiving is essential in many applications, in surveillance, for instance, also in television broadcasting, meteorology, and medical imaging, and archives are regularly accessed for decision making or simply to view a particular video segment. The exceedingly large and continually growing size of the archives precludes manual annotation to describe the video. It requires, instead, automatic means of describing and indexing the archives contents. By definition, motion is a fundamental dimension of video, which, therefore, can be used to extract descriptions which are characteristic of contents [22–25].

1.3 Variational Processing of Image Motion

The subject of this book is *image motion processing by variational methods*. Variational methods, rooted in physics and mechanics, but appearing in many other domains, such as statistics and control, address a problem from an optimization standpoint, i.e., they formulate it as the optimization of an objective function or functional. The methods of image motion analysis we describe in this book use calculus of variations to minimize (or maximize) an objective functional which transcribes all of the constraints that characterize the desired solution. For instance, optical flow estimation by the Horn and Schunck method [9] for an image sequence of domain Ω, $I : (x, y, t) \in \Omega \times]0, T[\mapsto I(x, y, t) \in \mathbf{R}^+$, where x, y are the spatial coordinates and t designates time, minimizes the following functional:

$$\mathscr{E}(u, v) = \int_\Omega (I_x u + I_y v + I_t)^2 dxdy + \lambda \int_\Omega (\|\nabla u\|^2 + \|\nabla v\|^2) dxdy \qquad (1.4)$$

where I_x, I_y, I_t are the image spatiotemporal derivatives, $\nabla u, \nabla v$ are the spatial gradients of optical flow and λ is a constant factor to weigh the contribution of the two terms of the functional. The first integral is a *data term* which evaluates the conformity of the motion field to the image spatiotemporal data. The other integral is a *regularization term* to bias the solution toward smooth motion fields. Data and smoothness terms are typical of the functionals we will investigate in this book. Using the calculus of variations [26], the minimization of Eq. (1.4), i.e, a solution of $\arg\min_{u,v} \mathscr{E}(u, v)$, can be obtained by solving the Euler-Lagrange equations corresponding to the functional, a topic we will address in Chap. 3.

The Horn and Schunck functional refers to the image domain Ω without attempting to divide it into specific subdomains characterized by specific descriptions. However, image motion analysis often brings in image segmentation naturally. For instance, motion detection is, by definition, the process of dividing the image domain into a foreground of moving objects and its complement, called the background, and three-dimensional interpretation of an image sequence requires segmenting in the image sequence the distinctly moving objects in space. For image motion analysis problems which involve also image segmentation, the use of closed regular plane curves to represent the segmentation boundaries is quite natural and useful. For example, if optical flow $W = (u, v)$ were estimated beforehand, motion detection can use the following functional:

$$\mathscr{E}(\gamma) = \int_{R_\gamma^c} \|W\| dxdy + \lambda \int_{R_\gamma} dxdy + \int_\gamma ds, \tag{1.5}$$

where γ is a closed regular plane curve which partitions the image domain Ω into two regions, R_γ, the interior of γ, to represent the foreground of moving objects, and its complement R_γ^c. Curve γ is the single variable to determine. The first two integrals are data terms. They assume, as explained in Chap. 4, that $\|W\| \geq \lambda$ for moving objects, i.e., they transcribe a thresholding scheme in the formulation. The last integral biases the solution toward smooth curves. Motion detection will seek the best partition, i.e., a curve that minimizes the functional. As a result, such a functional is called an *active curve functional* because the minimization equation with respect to γ is a curve evolution equation.

This book is organized according to the *four core subjects* of motion analysis: *Motion estimation, detection, tracking*, and *three-dimensional interpretation*. Each topic is covered in a dedicated chapter. The presentation is prefaced by Chap. 2 which gives brief descriptions and basic formulas related to curvature, Euler-Lagrange equations, unconstrained descent optimization, and level sets, all fundamental subjects or tools which the variational image motion processing methods in the book repeatedly use.

Chapter 3 covers *image motion estimation*, a topic that has been the focus of a considerable number of studies in computer vision. The purpose of this chapter is not to survey the vast literature but rather to offer a presentation of a few methods that would uncover, and explain to a certain extent, the fundamental concepts underlying image motion estimation by variational methods. The important concepts to expose include the optical flow constraint and its use in data terms of image motion estimation objective functionals; the use of the smoothness constraint in regularization terms; distinct views of the notion of motion discontinuity preservation which lead to distinct schemes of achieving it; the paradigm of joint motion estimation and segmentation by parametric representation; and, finally, the notions of mutiresolution and multigrid processing to treat large extent velocities. Processing of motion in stereoscopic image sequences will also be brought up. This reductive, concept-oriented presentation of image motion estimation will be complemented by a commented bibliography of recents studies which build upon the basic formulations we discuss and explain important computational aspects and details that are essential in any reasonable implementation.

Figure 1.2 shows two examples of motion estimation achievable by typical current methods. Figure 1.2a shows an image of a moving truck filmed by a stationary camera. The difficulty with the sequence is the lack of texture on the truck image and the relatively large image motion between successive frames of the sequence. The optical flow field computed by joint motion segmentation and parametric estimation [27], which accounts for these difficulties, is illustrated by the vector field superimposed on the image. Figure 1.2b is an image from a sequence showing a curled snake in movement. The raised head of the snake is immobile, and the rest of the body undergoes contortions. The motion is in one direction in the upper part of the body, in the opposite direction in the middle part, and almost stationary in the lower part, rendering a global representation of motion difficult. The results shown are also by joint motion segmentation and parametric estimation [27].

Motion detection by variational methods is covered in Chap. 4. Motion detection is currently the focus of intense research, particularly in key applications such as

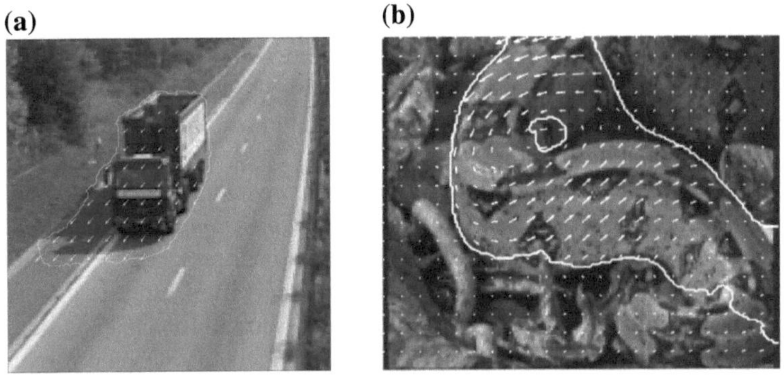

Fig. 1.2 **a** An image of a moving truck taken by a fixed camera and optical velocity vectors computed by variational joint motion estimation and segmentation; weak texture makes motion estimation difficult. **b** An image of snake contortions and computed optical velocity vectors. While the truck motion can be adequately modelled by affine motion, the contortions require an elaborate image motion representation

human activity analysis [17, 18, 28–32]. Its purpose being to determine the region in an image covered by moving objects, i.e., to divide an image into a foreground of moving objects and its complement, the use of an active curve is quite natural. The curve can be made to move according to characteristics of image motion in the background and foreground so as to bring it to coincide with the foreground boundary. We will describe both *region-based* and *boundary-based* active curve motion detection. Region-based detection uses motion information inside the moving curve and information outside. Boundary-based methods use information gathered along the curve, in which case the curve is called a *geodesic*. Most of the methods we will discuss will be for a static viewing system but the case of a moving system will also be considered. Figure 1.3 shows an example (from Chap. 4) of the kind of results variational formulations can achieve. In this particular example, the moving object (the image of the white car proceeding through the intersection) was detected by making a geodesic curve move to adhere to its boundary using optical flow contrast along the curve.

Chapter 5 addresses variational *motion tracking*, the process of following objects through an image sequence. The chapter starts with a brief presentation of the two major veins along which variational tracking has been investigated, namely discrete-time dynamic systems theory and energy minimization. This is followed with an extended review of a few current variational methods which use motion information about the targets they track. The methods include kernel-based tracking, supported by mean-shift optimization [33], distribution tracking [34], and temporal matching pursuit by active curves [35–37]. Figure 1.4 shows an example of the type of results which active contour variational methods can accomplish. In this particular instance, the method [36, 37] used both the photometric and shape profiles of the moving target simultaneously, which the variational formulation of tracking and the active contour representation of the target outline easily allowed to do.

(a) (b)

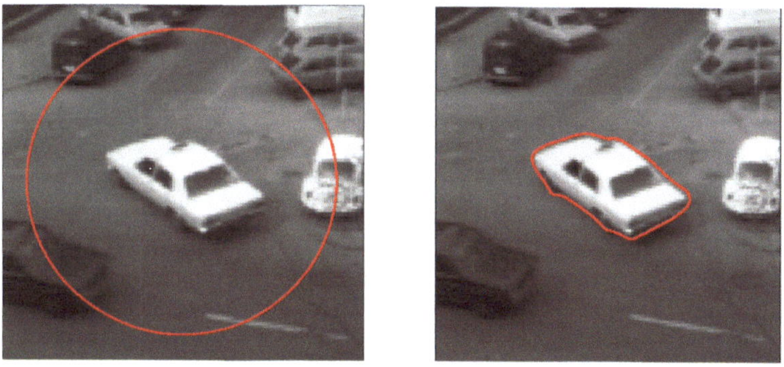

Fig. 1.3 Motion detection by minimizing a functional measuring the amount of optical flow contrast along a geodesic active curve: **a** The first of the two consecutive images used, and the initial curve superimposed and, **b** the final position of the curve. The active curve settled on the moving object boundary due to the strong image motion contrast along this boundary

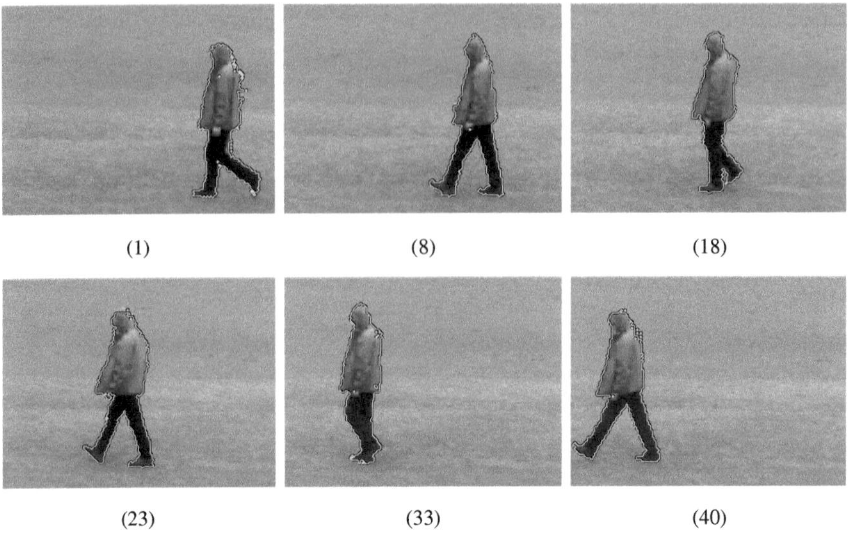

Fig. 1.4 Tracking of a walker in a grey scale sequence filmed by a static camera. This particular result was obtained using an active contour variational method driven both by the intensity profile of the walker image and the shape of its outline [36]. The shape information was used to constrain the active contour to retain a similar shape between consecutive frames of the sequence. The objective functional afforded a joint instantiation of the photometric and shape information to realize more accurate tracking

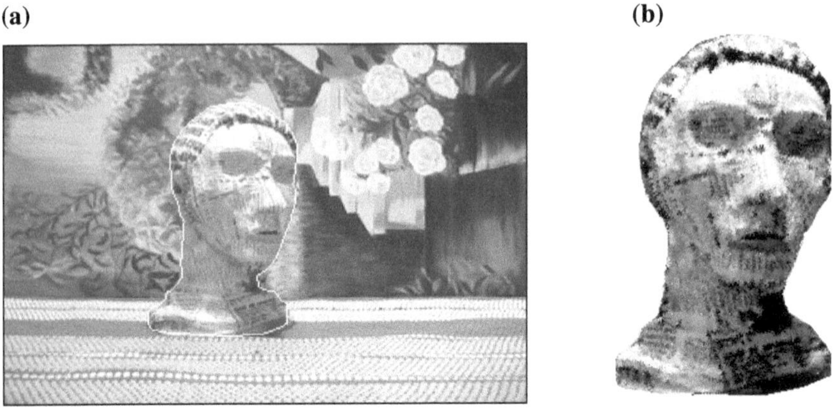

Fig. 1.5 Dense three-dimensional interpretation of optical flow by an active contour variational method [41]. The camera is static. The figurine moves in a static background. **a** The method was able to outline the moving figurine using 3D motion information which was concurrently recovered. **b** Displays a view, from a different angle, of the depth of the figurine constructed from the scheme's output

The purpose of Chap. 6 is *three-dimensional interpretation* of image motion, a long-standing topic of major importance in computer vision. Most of the related research is groundwork dealing with sparse sets of image points [38–40]. Dense variational interpretation of optical flow, which the chapter focuses on, has been the subject of much fewer later studies. The chapter discusses methods which use pre-computed optical flow and methods which compute and interpret it concurrently. It also distinguishes ego-motion, where only the viewing system moves, from general motion where objects and the viewing system can move. Most of the methods described assume rigid objects in the environment but scene flow estimation to recover the moving surfaces relative 3D velocity field without making such a hypothesis is also addressed. Figure 1.5 show an example of the type of results that can be obtained by active contour variational methods. Such methods realize joint motion-based image segmentation and 3D motion estimation [42].

References

1. G. Wald, Eye and camera. Sci. Am. **8** (August), 32–41 (1950)
2. K. Nakayama, Biological image motion processing: a survey. Vision Res. **25**, 625–660 (1985)
3. H. von Helmholtz, Handbook of Physiological Optics, Vol. 3. Verlag von Leopold Voss, Editor: J.P.C. Southall, Optical Society of America, 1910; 1925 for the English version
4. J.J. Gibson, *The Perception of the Visual World* (Houghton Mifflin, Boston, 1950)
5. H. Wallach, D.N. O'Connell, The kinetic depth effect. J. Exp. Psychol. **45**, 205–217 (1953)
6. E.J. Gibson, J.J. Gibson, O.W. Smith, H. Flock, Motion parallax as a determinant of depth. J. Exp. Psychol. **58**, 40–51 (1959)
7. S. Ullman, The interpretation of structure from motion. Proc. R. Soc. Lond. B **203**, 405–426 (1979)
8. B.J. Rogers, M. Graham, Motion parallax as an independent cue for depth perception. Vision Res. **8**, 125–134 (1979)
9. B. Horn, B. Schunck, Determining optical flow. Artif. Intell. **17**, 185–203 (1981)
10. H.C. Longuet-Higgins, K. Prazdny, The interpretation of a moving retinal image. Proc. R. Soc. Lond. B **208**, 385–397 (1980)
11. R.M. Haralick, L.G. Shapiro, *Computer and Robot Vision* (Addison Wesley, Reading, 1992)
12. X. Armangue, H. Araujo, J. Salvi, A review on egomotion by means of differential epipolar geometry applied to the movement of a mobile robot. Pattern Recogn. **36**(12), 2927–2944 (2003)
13. N.E. Mortensen, *Progress in Autonomous Robot Research* (Nova Science Publishers, New York, 2008)
14. A. Crétual, F. Chaumette, Visual servoing based on image motion. Int. J. Robot. Res. **20**(11), 857–877 (2001)
15. F. Chaumette, S. Hutchinson, Visual servo control, part i: Basic approaches. IEEE Robot. Autom. Mag. **13**(4), 82–90 (2006)
16. F. Chaumette, S. Hutchinson, Visual servo control, part ii: Advanced approaches. IEEE Robot. Autom. Mag. **14**(1), 109–118 (2007)
17. T.B. Moeslund, A. Hilton, V. Krüger, A survey of advances in vision-based human motion capture and analysis. Comput. Vis. Image Underst. **104**(2–3), 90–126 (2006)
18. R. Poppe, Vision-based human motion analysis: an overview. Comput. Vis. Image Underst. **108**(1–2), 4–18 (2007)
19. J.K. Aggarwal, M. Ryoo, Human activity analysis: A review. ACM Comput. Surv. **43**(3), 16:1–16:43 (2011). http://doi.acm.org/10.1145/1922649.1922653

20. W. Hu, T. Tan, L. Wang, S. Maybank, A survey on visual surveillance of object motion and behaviors. IEEE Trans. Syst. Man Cybern. **34**, 334–352 (2004)
21. L. Yu, J.-p. Wang, Review of the current and future technologies for video compression. J. Zhejiang Univ. Sci. C, **11**, 1–13 (2010). 10.1631/jzus.C0910684. http://dx.doi.org/10.1631/jzus.C0910684
22. R. Brunelli, O. Mich, C. Modena, A survey on the automatic indexing of video data. J. Vis. Commun. Image Represent. **10**(2), 78–112 (1999)
23. R. Fablet, P. Bouthemy, P. Perez, Nonparametric motion characterization using causal probabilistic models for video indexing and retrieval. IEEE Trans. Image Process. **11**(4), 393–407 (2002)
24. G. Piriou, P. Bouthemy, J.-F. Yao, Learned probabilistic image motion models for event detection in videos, in *IEEE International Conference on Computer Vision*, vol. 4, 2004, pp. 207–210
25. G. Piriou, P. Bouthemy, J.-F. Yao, Recognition of dynamic video contents with global probabilistic models of visual motion. IEEE Trans. Image Process. **15**(11), 3417–3430 2006
26. R. Weinstock, *Calculus of Variations* (Dover, New York, 1974)
27. C. Vazquez, A. Mitiche, R. Laganiere, Joint segmentation and parametric estimation of image motion by curve evolution and level sets. IEEE Trans. Pattern Anal. Mach. Intell. **28**(5), 782–793 (2006)
28. L. Wang, W. Hu, T. Tan, Recent developments in human motion analysis. Pattern Recognit. **36**(3), 585–601 (2003)
29. C. Sminchisescu, 3D human motion analysis in monocular video techniques and challenges, in *AVSS*, 2006, p. 76
30. X. Ji, H. Liu, Advances in view-invariant human motion analysis: A review. IEEE Trans. Syst. Man Cybern. C **40**(1), 13–24 (2010)
31. Z. Sun, G. Bebis, R. Miller, On-road vehicle detection: a review. IEEE Trans. Pattern Anal. Mach. Intell. **28**(5), 694–711 (2006)
32. M. Enzweiler, D.M. Gavrila, Monocular pedestrian detection: survey and experiments. IEEE Trans. Pattern Anal. Mach. Intell. **31**(12), 2179–2195 (2009)
33. D. Comaniciu, V. Ramesh, P. Meer, Kernel-based object tracking. IEEE Trans. Pattern Anal. Mach. Intell. **25**(5), 564–575 (2003)
34. D. Freedman, T. Zhang, Active contours for tracking distributions. IEEE Trans. Image Process. **13**(4), 518–526 (2004)
35. A. Mansouri, Region tracking via level set pdes without motion computation. IEEE Trans. Pattern Anal. Mach. Intell. **24**(7), 947–961 (2002)
36. M. Ben Salah, Fonctions noyaux et a priori de forme pour la segmentation d'images et le suivi d'objets, Ph.D. dissertation, Institut national de la recherche scientifique, INRS-EMT, 2011
37. M. Ben Salah, A. Mitiche, Model-free, occlusion accommodating active contour tracking. ISRN Artif. Intell. **2012**, Article ID 672084, 15 (2012)
38. J.K. Aggarwal, N. Nandhakumar, On the computation of motion from a sequence of images: a review. Proc. IEEE **76**, 917–935 (1988)
39. A. Mitiche, *Computational Analysis of Visual Motion* (Plenum Press, New York, 1994)
40. T. Huang, A. Netravali, Motion and structure from feature correspondences: a review. Proc. IEEE **82**, 252–268 (1994)
41. H. Sekkati, A. Mitiche, Joint optical flow estimation, segmentation, and 3D interpretation with level sets. Comput. Vis. Image Underst. **103**(2), 89–100 (2006)
42. A. Mitiche, I. Ben Ayed, *Variational and Level Set Methods in Image Segmentation* (Springer, New York, 2010)

Chapter 2
Background Preliminaries

In this preliminary chapter we will give definitions, descriptions, and formulas, concerning curvature, Euler-Lagrange equations, unconstrained descent optimization, and level sets, all fundamental topics and tools underlying the variational methods of motion analysis described in the subsequent chapters.

2.1 Curvature

Active curve objective functionals of the type we investigate in this book often contain a term which measures the length of a regular closed plane curve, or which integrates a scalar function along such a curve. These terms produce curvature in the objective functional minimization equations. In this section we review some basic facts about curvature of plane curves.

2.1.1 Curvature of a Parametric Curve

A *parametrized differentiable plane curve* is a differentiable map $\mathbf{c} : J \rightarrow \mathbb{R}^2$, from an open interval $J \subset \mathbb{R}$ into $\mathbb{R}^2$, i.e., a correspondence which maps each $r \in J$ to a point $(x(r), y(r))$ of the plane in such a way that the coordinate functions $x(r)$ and $y(r)$ are differentiable. The vector $\mathbf{c}'(r) = (x'(r), y'(r))$ of first derivatives of the coordinate functions is the *tangent vector*, or *velocity vector*, of the curve $\mathbf{c}$ at r. A *regular* curve is a differentiable curve for which $\mathbf{c}'(r) \neq 0$ for all r. For $r \in J$, the *arc length* of a regular curve $\mathbf{c}$ from a point r_0 is defined as the function:

$$s(r) = \int_{r_0}^{r} \|\mathbf{c}'(z)\| dz \qquad (2.1)$$

A. Mitiche and J. K. Aggarwal, *Computer Vision Analysis of Image Motion by Variational Methods*, Springer Topics in Signal Processing 10, DOI: 10.1007/978-3-319-00711-3_2,

Since $\mathbf{c}'(r) \neq \mathbf{0}$, the arc length function is differentiable and

$$\frac{ds}{dr}(r) = \|\mathbf{c}'(r)\| = \left((x'(r))^2 + (y'(r))^2\right)^{1/2}. \tag{2.2}$$

When r is itself arc length, i.e., $s(r) = r$, then we have $ds/dr(r) = 1$ and, therefore, $\|\mathbf{c}'(r)\| = 1$, prompting the definition: A curve $\mathbf{c} : J \to \mathbb{R}^2$ is said to be parametrized by arc length s if $\|\mathbf{c}'(s)\| = 1 \;\; \forall s \in J$,

Let $\mathbf{c} : r \in J \to (x(r), y(r)) \in \mathbb{R}^2$ be a regular parametrized plane curve, not necessarily by arc length. The parametrization defines two possible orientations of the curve: The orientation along which the parameter grows and the opposite orientation. Let $\mathbf{t}$ be the unit tangent vector of $\mathbf{c}$ associated with the orientation of growing curve parameter (Fig. 2.1). We have:

$$\mathbf{t} = \left(\frac{x'}{\left((x')^2 + (y')^2\right)^{1/2}}, \frac{y'}{\left((x')^2 + (y')^2\right)^{1/2}} \right), \tag{2.3}$$

where the prime symbol designates the derivative with respect to parameter r.

Define now the unit normal vector $\mathbf{n}$ by requiring the basis $(\mathbf{t}, \mathbf{n})$ to have the same orientation as the natural basis (e_1, e_2) of $\mathbb{R}^2$ (Fig. 2.1) and then *define* curvature κ by [1]:

$$\frac{d\mathbf{t}}{ds} = \kappa \mathbf{n}, \tag{2.4}$$

where s is arc length. This definition gives a sign to the curvature, i.e., it can be positive or negative depending on the point of evaluation. This can be quickly verified graphically by drawing a figure such as Fig. 2.1. Changing the orientation of the curve or of $\mathbb{R}^2$ will change the sign of the curvature.

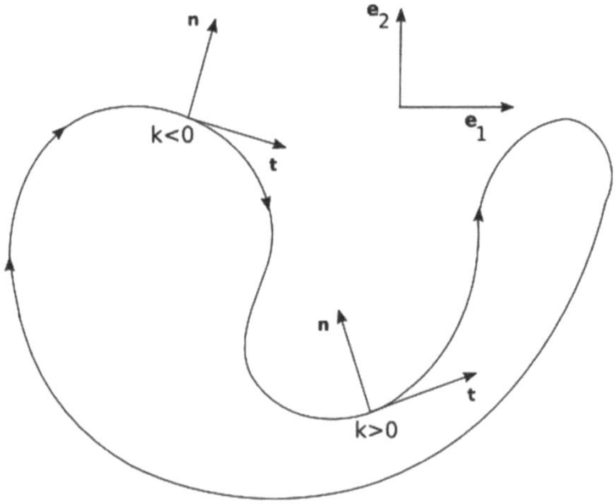

Fig. 2.1 For plane curves, curvature can be given a sign as follows: Let $\mathbf{t}$ be the unit tangent vector and define the unit normal vector $\mathbf{n}$ by requiring the basis $(\mathbf{t}, \mathbf{n})$ to have the same orientation as the natural basis of $\mathbb{R}^2$. Then define curvature κ by $d\mathbf{t}/ds = \kappa \mathbf{n}$, where s is the arc length parameter. The sign of the curvature changes if the orientation of either the curve or of the normal changes

Derivation of the unit tangent vector, with respect to parameter r, gives:

$$\begin{aligned}\frac{d\mathbf{t}}{dr} &= \left(-y'\frac{x'y''-y'x''}{\left((x')^2+(y')^2\right)^{3/2}}\,,\; x'\frac{x'y''-y'x''}{\left((x')^2+(y')^2\right)^{3/2}}\right)\\ &= \frac{x'y''-y'x''}{\left((x')^2+(y')^2\right)^{3/2}}\left(-y'\,,\; x'\right).\end{aligned} \tag{2.5}$$

Using Eq. (2.2), noting that dr/ds is the inverse of $ds/dr \neq 0$, i.e., $dr/ds = 1/\|\mathbf{c}'(r)\| = \left((x'(r))^2+(y'(r))^2\right)^{-1/2}$, we have:

$$\begin{aligned}\frac{d\mathbf{t}}{ds} = \frac{d\mathbf{t}}{dr}\frac{dr}{ds} &= \frac{x'y''-y'x''}{\left((x')^2+(y')^2\right)^{3/2}}\left(-\frac{y'}{\left((x')^2+(y')^2\right)^{1/2}}\,,\; \frac{x'}{\left((x')^2+(y')^2\right)^{1/2}}\right)\\ &= \frac{x'y''-y'x''}{\left((x')^2+(y')^2\right)^{3/2}}\mathbf{n}\end{aligned} \tag{2.6}$$

Comparing Eq. (2.6) to Eq. (2.4), we have the following parametrization independent expression of curvature:

$$\kappa = \frac{x'y''-y'x''}{\left((x')^2+(y')^2\right)^{3/2}} \tag{2.7}$$

2.1.2 Curvature of an Implicit Curve

Assume that the level set $\{(x, y)|\phi(x, y) = 0\}$ defines a differentiable parametric curve $\mathbf{c} : r \to \mathbf{c}(r) = (x(r), y(r))$. Then by the chain rule of differentiation:

$$\frac{d}{dr}\phi(\mathbf{c}(r)) = \nabla\phi \cdot \mathbf{v} = 0, \tag{2.8}$$

where $\mathbf{v} = \mathbf{c}'$ is the tangent vector of $\mathbf{c}$. This shows that the gradient of ϕ is perpendicular to the level set curve. Let $\mathbf{n}$ be the unit vector normal to $\mathbf{c}$ defined by:

$$\mathbf{n} = \frac{\nabla\phi}{\|\nabla\phi\|} = \left(\frac{\phi_x}{\left(\phi_x^2+\phi_y^2\right)^{1/2}}, \frac{\phi_y}{\left(\phi_x^2+\phi_y^2\right)^{1/2}}\right) \tag{2.9}$$

Let the unit tangent vector $\mathbf{t}$ be defined by requiring the basis $\{\mathbf{t}, \mathbf{n}\}$ to have the same orientation as the natural basis $\{e_1, e_2\}$ of $\mathbb{R}^2$, we have:

$$\mathbf{t} = \left(\frac{\phi_y}{\left(\phi_x^2 + \phi_y^2\right)^{1/2}}, \frac{-\phi_x}{\left(\phi_x^2 + \phi_y^2\right)^{1/2}} \right), \tag{2.10}$$

Curvature can then be defined by Eq. (2.4).

Let s designate arc length. Using the chain rule of differentiation we can write:

$$\frac{d\mathbf{t}}{ds} = \frac{\partial \mathbf{t}}{\partial x}\frac{dx}{ds} + \frac{\partial \mathbf{t}}{\partial y}\frac{dy}{ds}. \tag{2.11}$$

In this expression, $\left(\frac{dx}{ds}, \frac{dy}{ds}\right) = \mathbf{t}$; therefore, substitution in Eq. (2.11) of expression Eq. (2.10) of $\mathbf{t}$ gives:

$$\frac{d\mathbf{t}}{ds} = \frac{1}{\|\nabla\phi\|}\left(\phi_y \frac{\partial \mathbf{t}}{\partial x} - \phi_x \frac{\partial \mathbf{t}}{\partial y}\right). \tag{2.12}$$

The partial derivative with respect to x of the first component of $\mathbf{t}$ evaluates as follows:

$$\begin{aligned} \frac{\partial}{\partial x}\left(\frac{\phi_y}{\left(\phi_x^2+\phi_y^2\right)^{1/2}}\right) &= \frac{\phi_{xy}}{\left(\phi_x^2+\phi_y^2\right)^{1/2}} - \frac{\phi_y(\phi_x\phi_{xx}+\phi_y\phi_{xy})}{(\phi_x^2+\phi_y^2)^{3/2}} \\ &= \frac{\phi_{xy}(\phi_x^2+\phi_y^2) - \phi_y(\phi_x\phi_{xx}+\phi_y\phi_{xy})}{(\phi_x^2+\phi_y^2)^{3/2}} \\ &= \frac{\phi_{xy}\phi_x^2 - \phi_x\phi_y\phi_{xx}}{(\phi_x^2+\phi_y^2)^{3/2}} \end{aligned} \tag{2.13}$$

Similarly, we determine that the partial derivative with respect to y of the first component of $\mathbf{t}$ is given by:

$$\frac{\partial}{\partial y}\left(\frac{\phi_y}{\left(\phi_x^2+\phi_y^2\right)^{1/2}}\right) = \frac{\phi_{yy}\phi_x^2 - \phi_x\phi_y\phi_{xy}}{(\phi_x^2+\phi_y^2)^{3/2}}. \tag{2.14}$$

Substitution of Eqs. (2.13) and (2.14) back in Eq. (2.12) gives the derivative with respect to s of the first component of $\mathbf{t}$:

$$\frac{dt_1}{ds} = -\frac{\phi_x}{\|\nabla\phi\|}\frac{\phi_{xx}\phi_y^2 - 2\phi_x\phi_y\phi_{xy} + \phi_{yy}\phi_x^2}{(\phi_x^2+\phi_y^2)^{3/2}} \tag{2.15}$$

We proceed in the same manner to obtain the partial derivatives with respect to x and y of the second component of the tangent vector $\mathbf{t}$:

$$\frac{\partial}{\partial x}\left(\frac{-\phi_x}{\left(\phi_x^2+\phi_y^2\right)^{1/2}}\right)=\frac{\phi_x\phi_y\phi_{xy}-\phi_{xx}\phi_y^2}{(\phi_x^2+\phi_y^2)^{3/2}} \tag{2.16}$$

$$\frac{\partial}{\partial y}\left(\frac{-\phi_x}{\left(\phi_x^2+\phi_y^2\right)^{1/2}}\right)=\frac{\phi_x\phi_y\phi_{yy}-\phi_{xy}\phi_y^2}{(\phi_x^2+\phi_y^2)^{3/2}}, \tag{2.17}$$

which give the derivative with respect to s of the second component of **t** by substitution into Eq. (2.12):

$$\frac{dt_2}{ds}=-\frac{\phi_y}{\|\nabla\phi\|}\frac{\phi_{xx}\phi_y^2-2\phi_x\phi_y\phi_{xy}+\phi_{yy}\phi_x^2}{(\phi_x^2+\phi_y^2)^{3/2}} \tag{2.18}$$

Putting together Eq. (2.15) and Eq. (2.18) gives the desired equation:

$$\frac{d\mathbf{t}}{ds}=-\frac{\phi_{xx}\phi_y^2-2\phi_x\phi_y\phi_{xy}+\phi_{yy}\phi_x^2}{(\phi_x^2+\phi_y^2)^{3/2}}\mathbf{n} \tag{2.19}$$

Comparing with Eq. (2.4), we have the expression of curvature:

$$\kappa=-\frac{\phi_{xx}\phi_y^2-2\phi_x\phi_y\phi_{xy}+\phi_{yy}\phi_x^2}{(\phi_x^2+\phi_y^2)^{3/2}} \tag{2.20}$$

Curvature can also be expressed as, with our choice of **n** and **t**:

$$\kappa=-\mathrm{div}\left(\frac{\nabla\phi}{\|\nabla\phi\|}\right), \tag{2.21}$$

which can be proved by expanding the righthand side:

$$-\,\mathrm{div}\left(\frac{\nabla\phi}{\|\nabla\phi\|}\right) \tag{2.22}$$

$$=-\frac{\partial}{\partial x}\left(\frac{\phi_x}{\left(\phi_x^2+\phi_y^2\right)^{1/2}}\right)-\frac{\partial}{\partial y}\left(\frac{\phi_y}{\left(\phi_x^2+\phi_y^2\right)^{1/2}}\right) \tag{2.23}$$

$$=-\frac{\phi_{xx}}{\left(\phi_x^2+\phi_y^2\right)^{1/2}}+\frac{\phi_x(\phi_x\phi_{xx}+\phi_y\phi_{xy})}{(\phi_x^2+\phi_y^2)^{3/2}}-\frac{\phi_{yy}}{\left(\phi_x^2+\phi_y^2\right)^{1/2}}+\frac{\phi_y(\phi_x\phi_{xy}+\phi_y\phi_{yy})}{(\phi_x^2+\phi_y^2)^{3/2}} \tag{2.24}$$

$$= -\frac{\phi_{xx}(\phi_x^2+\phi_y^2) - \phi_x(\phi_x\phi_{xx}+\phi_y\phi_{xy}) + \phi_{yy}(\phi_x^2+\phi_y^2) - \phi_y(\phi_x\phi_{xy}+\phi_y\phi_{yy})}{(\phi_x^2+\phi_y^2)^{3/2}} \tag{2.25}$$

$$= -\frac{\phi_{xx}\phi_y^2 - 2\phi_x\phi_y\phi_{xy} + \phi_{yy}\phi_x^2}{(\phi_x^2+\phi_y^2)^{3/2}}, \tag{2.26}$$

which is the same expression as given in Eq. (2.20).[1]

2.2 Euler-Lagrange Equations

The active curve objective functionals we will investigate in this book are minimized by solving the corresponding Euler-Lagrange equations.[2] Here following is a review of the basic formulas we will be using, concerning both definite integrals and variable domain integrals. The variable domain integrals we will treat include curve length integrals of a closed regular plane curve, integrals of a scalar function along a closed regular plane curve, or over a closed regular surface in $\mathbb{R}^3$, as well as integrals over bounded regions in $\mathbb{R}^2$ and $\mathbb{R}^3$.

2.2.1 *Definite Integrals*

The purpose in this section is to provide succinct derivations of the basic Euler-Lagrange differential equations for definite integrals. We will follow the presentation of R. Weinstock [2] which requires only basic results in vector calculus [3]. We will first develop the Euler-Lagrange equation corresponding to an integral involving a real function of a real variable:

$$\mathscr{E}(u) = \int_{x_1}^{x_2} g(x, u, u')dx, \tag{2.27}$$

where the endpoints x_1 and x_2 are given real numbers; $u = u(x)$ is a twice differentiable real function; $u' = \frac{du}{dx}$; and g is a function twice differentiable with respect to any of its three arguments, x, u, and u'.

Assume that there exists a twice-differentiable function u satisfying the boundary conditions $u(x_1) = u_1$ and $u(x_2) = u_2$ and which minimizes the integral Eq. (2.27). We want to determine the differential equation which this minimizer u must satisfy. To do this, let $\eta(x)$ be an arbitrary differentiable function which satisfies the endpoint

---

[1] Note that we could have defined the unit normal of the implicit curve as $\mathbf{n} = -\nabla\phi/\|\nabla\phi\|$ instead of $\mathbf{n} = \nabla\phi/\|\nabla\phi\|$ as in Eq. (2.9), in which case curvature would change sign, i.e., it would have the expression in Eq. (2.20) but without the minus sign. At the same time it would be written $\kappa = \text{div}\,(\nabla\phi/\|\nabla\phi\|)$ rather than with the minus sign as in Eq. (2.21).

[2] The discussions apply as well to the maximization of similar functionals.

conditions $\eta(x_1) = \eta(x_2) = 0$, and define the following one-parameter family of functions U indexed by parameter $\varepsilon \in \mathbb{R}$:

$$U(x, \varepsilon) = u(x) + \varepsilon\eta(x) \tag{2.28}$$

All functions U in this family have the same endpoints as u, i.e., $U(x_1, \varepsilon) = u_1$ and $U(x_2, \varepsilon) = u_2$ for all ε. The minimizer u is the member of the family corresponding to $\varepsilon = 0$, i.e., $U(x, 0) = u(x)$. By definition, there is a neighborhood $\mathscr{U}$ of u where the integral is a minimum at u, i.e., $\mathscr{E}(u) \leq \mathscr{E}(y)\ \forall y \in \mathscr{U}$. We can choose ε in a small enough interval J so that the functions so defined by Eq. (2.28) fall in this neighborhood for all $\varepsilon \in J$. In this case the following integral function of ε:

$$E(\varepsilon) = \int_{x_1}^{x_2} g(x, U(x, \varepsilon), U'(x, \varepsilon))dx, \tag{2.29}$$

with

$$U' = \frac{dU}{dx} = u' + \varepsilon\eta', \tag{2.30}$$

is minimized with respect to ε for $\varepsilon = 0$ and, therefore:

$$\frac{dE}{d\varepsilon}(0) = 0 \tag{2.31}$$

Differentiation under the integral sign (Sect. 2.3) of Eq. (2.29) with respect to parameter ε gives:

$$\begin{aligned} \frac{dE}{d\varepsilon}(\varepsilon) &= \int_{x_1}^{x_2} \left(\frac{\partial g}{\partial U}\frac{\partial U}{\partial \varepsilon} + \frac{\partial g}{\partial U'}\frac{\partial U'}{\partial \varepsilon} \right) dx \\ &= \int_{x_1}^{x_2} \left(\frac{\partial g}{\partial U}\eta + \frac{\partial g}{\partial U'}\eta' \right) dx \end{aligned} \tag{2.32}$$

The necessary condition Eq. (2.31) is then written as:

$$\frac{dE}{d\varepsilon}(0) = \int_{x_1}^{x_2} \left(\frac{\partial g}{\partial u}\eta + \frac{\partial g}{\partial u'}\eta' \right) dx = 0 \tag{2.33}$$

Integration by parts of the second term of the integrand gives:

$$\frac{dE}{d\varepsilon}(0) = \int_{x_1}^{x_2} \left[\frac{\partial g}{\partial u} - \frac{d}{dx}\left(\frac{\partial g}{\partial u'} \right) \right] \eta \, dx = 0. \tag{2.34}$$

This equation must hold for all η. Therefore, we have the *Euler-Lagrange equation* which the minimizer u of $\mathscr{E}$ must satisfy:

$$\frac{\partial g}{\partial u} - \frac{d}{dx}\left(\frac{\partial g}{\partial u'}\right) = 0. \tag{2.35}$$

Several Dependent Variables

When the integral involves several dependent real variables $u(x), v(x), \ldots, w(x)$:

$$\mathscr{E}(u) = \int_{x_1}^{x_2} g(x, u, v, \ldots, w, u', v', \ldots, w')dx, \tag{2.36}$$

similar developments yield one Euler-Lagrange equation for each dependent variable:

$$\begin{aligned} \frac{\partial g}{\partial u} - \frac{d}{dx}\left(\frac{\partial g}{\partial u'}\right) &= 0 \\ \frac{\partial g}{\partial v} - \frac{d}{dx}\left(\frac{\partial g}{\partial v'}\right) &= 0 \\ &\cdots \\ \frac{\partial g}{\partial w} - \frac{d}{dx}\left(\frac{\partial g}{\partial w'}\right) &= 0 \end{aligned} \tag{2.37}$$

Several Independent Variables

In subsequent chapters, we will encounter integrals involving scalar functions of two independent variables. Consider an integral of the form:

$$\mathscr{E}(w) = \int_R g(x, y, w, w_x, w_y)dxdy, \tag{2.38}$$

where R is a bounded region of $\mathbb{R}^2$ the boundary ∂R of which is a regular closed plane curve; $w = w(x, y)$ assumes some prescribed values at all points on ∂R; w_x and w_y are the partial derivatives of w; and g is twice continuously differentiable with respect to its arguments. To determine the differential equation which a minimizing function $w(x, y)$ must satisfy, we proceed at first as we did with functions of a single real variable, namely: we consider the following family of functions indexed by real parameter ε:

$$W(x, y, \varepsilon) = w(x, y) + \varepsilon\eta(x, y), \tag{2.39}$$

where η is an arbitrary continuously differentiable real function such that $\eta(x, y) = 0$ on ∂R, so that functions W all have the same boundary values. Then we form the integral function of ε:

$$E(\varepsilon) = \int_R g(x, y, W(x, y, \varepsilon), W_x(x, y, \varepsilon), W_y(x, y, \varepsilon))dxdy, \tag{2.40}$$

and remark that

$$\frac{dE}{d\varepsilon}(0) = 0, \tag{2.41}$$

which would give:

$$\frac{dE}{d\varepsilon}(0) = \int_R \left(\frac{\partial g}{\partial w}\eta + \frac{\partial g}{\partial w_x}\frac{\partial \eta}{\partial x} + \frac{\partial g}{\partial w_y}\frac{\partial \eta}{\partial y} \right) dxdy = 0. \tag{2.42}$$

To continue the derivation we need to apply the Green's theorem to the integrals corresponding to the last two terms of the integrand. We recall the theorem in its most usual form: Let $P(x, y)$ and $Q(x, y)$ be real functions with continuous first partial derivatives in a region R of the plane bounded by a regular closed curve. Then:

$$\int_R \left(\frac{\partial P}{\partial x} + \frac{\partial Q}{\partial y} \right) dxdy = \int_{\partial R} (Pdy - Qdx). \tag{2.43}$$

We will use the integration by parts expression of this theorem, obtained by setting $P = \eta G$ and $Q = \eta F$ [2]:

$$\int_R \left(G\frac{\partial \eta}{\partial x} + F\frac{\partial \eta}{\partial y} \right) dxdy = -\int_R \eta \left(\frac{\partial G}{\partial x} + \frac{\partial F}{\partial y} \right) dxdy + \int_{\partial R} \eta(Gdy - Fdx). \tag{2.44}$$

In our case, $G = \frac{\partial g}{\partial w_x}$; $F = \frac{\partial g}{\partial w_y}$, and the second integral on the righthand side of Eq. (2.44) is zero because $\eta = 0$ on ∂R. Applying this to Eq. (2.42) gives:

$$\int_R \eta \left[\frac{\partial g}{\partial w} - \frac{\partial}{\partial x}\left(\frac{\partial g}{\partial w_x} \right) - \frac{\partial}{\partial y}\left(\frac{\partial g}{\partial w_y} \right) \right] dxdy = 0. \tag{2.45}$$

This is an equation which must be satisfied for all η, leading to the desired Euler-Lagrange equation:

$$\frac{\partial g}{\partial w} - \frac{\partial}{\partial x}\left(\frac{\partial g}{\partial w_x} \right) - \frac{\partial}{\partial y}\left(\frac{\partial g}{\partial w_y} \right) = 0. \tag{2.46}$$

In the general case of more than two independent variables $x, y, \ldots, z$, Eq. (2.46) generalizes to:

$$\frac{\partial g}{\partial w} - \frac{\partial}{\partial x}\left(\frac{\partial g}{\partial w_x} \right) - \frac{\partial}{\partial y}\left(\frac{\partial g}{\partial w_y} \right) - \cdots - \frac{\partial}{\partial z}\left(\frac{\partial g}{\partial w_z} \right) = 0. \tag{2.47}$$

Example: Let $I : (x, y, t) \in \Omega \times]0, T[\mapsto I(x, y, t) \in \mathbb{R}^+$ be an image sequence and consider the Horn and Schunck optical flow estimation functional:

$$\mathscr{E}(u,v) = \int_{\Omega} (I_x u + I_y v + I_t)^2 dxdy + \lambda \int_{\Omega} (\|\nabla u\|^2 + \|\nabla v\|^2) dxdy$$

where I_x, I_y, I_t are the image spatiotemporal derivatives, $\nabla u, \nabla v$ are the spatial gradients of the optical flow coordinates u, v, and λ is a constant factor to weigh the contribution of the two terms in the objective functional. There are two independent variables, namely the image coordinates x, y and two dependent variables, namely the functions $u(x, y), v(x, y)$. Therefore, we will have two equations, one for u and one for v. We apply Eq. (2.46) to each of u and v with $g(x, y, u, v, u_x, u_y, v_x, v_y) = (I_x u + I_y v + I_t)^2 + \lambda(u_x^2 + u_y^2 + v_x^2 + v_y^2)$ which immediately gives:

$$\begin{aligned} I_x(I_x u + I_y v + I_t) - \lambda \nabla^2 u = 0 \\ I_y(I_x u + I_y v + I_t) - \lambda \nabla^2 v = 0, \end{aligned} \tag{2.48}$$

where $\nabla^2 = \partial^2/\partial x^2 + \partial^2/\partial y^2$ is the Laplacian operator.

Functional derivative: As frequently done in the computer vision literature, we will refer to the lefthand side of the Euler-Lagrange equation of an integral $\mathscr{E}$ corresponding to a dependent real variable u as the *functional derivative* of $\mathscr{E}$ with respect to u, with the notation $\frac{d\mathscr{E}}{du}$ when u is the single argument of $\mathscr{E}$ and $\frac{\partial \mathscr{E}}{\partial u}$ when $\mathscr{E}$ has several arguments.

2.2.2 Variable Domain of Integration

We will derive the functional derivative of functionals which are common in image motion analysis and image segmentation by active contours, and which appear throughout this book, namely integrals over regions enclosed by closed regular plane curves and path integrals of scalar functions over such curves. We will also derive the functional derivative for surface and volume integrals.

Region Integral of a Scalar Function

Let R_γ be the interior of a closed regular plane curve parametrized by arc length, $\gamma : s \in [0, l] \rightarrow (x(s), y(s)) \in \mathbb{R}$. The segmentation functionals we will encounter in this book typically contain a term of the form:

$$\mathscr{E}(\gamma) = \int_{R_\gamma} f(x, y) dxdy, \tag{2.49}$$

where f is a scalar function, i.e., independent of γ. The functional depends on γ via its domain of integration which is a function of γ.

To determine the Euler-Lagrange equation corresponding to the minimization of Eq. (2.49) with respect to γ (we assume that the problem is to minimize $\mathscr{E}$ but, of course, the discussion applies to maximization as well), the functional is first transformed into a simple integral as follows using Green's theorem [3]. Let

$$P(x, y) = -\frac{1}{2} \int_0^y f(x, z)dz \tag{2.50}$$

and

$$Q(x, y) = \frac{1}{2} \int_0^x f(z, y)dz \tag{2.51}$$

According to Green's theorem we have:

$$\int_{R_\gamma} \left(\frac{\partial Q}{\partial x} - \frac{\partial P}{\partial y} \right) dxdy = \int_\gamma Pdx + Qdy \tag{2.52}$$

Since $\frac{\partial Q}{\partial x} - \frac{\partial P}{\partial y} = f(x, y)$ we get:

$$\int_{R_\gamma} f(x, y)dxdy = \int_\gamma Pdx + Qdy = \int_0^l \left(Px' + Qy'\right) ds, \tag{2.53}$$

where $x' = \frac{dx}{ds}$ and $y' = \frac{dy}{ds}$. Applying Eq. (2.37) to the last integral in Eq. (2.53), i.e., using:

$$g(s, x, y, x', y') = P(x(s), y(s))x'(s) + Q(x(s), y(s))y'(s), \tag{2.54}$$

we get two equations, one for each component function of γ:

$$\begin{aligned} \frac{\partial \mathscr{E}}{\partial x} &= \frac{\partial g}{\partial x} - \frac{d}{ds}\left(\frac{\partial g}{\partial x'}\right) = \left(\frac{\partial Q}{\partial x} - \frac{\partial P}{\partial y}\right) y' = fy' \\ \frac{\partial \mathscr{E}}{\partial y} &= \frac{\partial g}{\partial y} - \frac{d}{ds}\left(\frac{\partial g}{\partial y'}\right) = \left(-\frac{\partial Q}{\partial x} + \frac{\partial P}{\partial y}\right) x' = -fx'. \end{aligned} \tag{2.55}$$

The Green's theorem expression in Eq. (2.52) assumes that curve γ is oriented counter clockwise [3]. With this orientation and since we are using the arc length parametrization, the outward unit normal $\mathbf{n}$ to γ is $\mathbf{n} = (y', -x')$. Therefore, the functional derivatives in Eq. (2.55) can be written in vector form as:

$$\frac{\partial \mathscr{E}}{\partial \gamma} = f\mathbf{n} \tag{2.56}$$

The Length Integral (Two Dimensions)

Another functional which very often appears in the motion analysis formulations in the book is the curve length functional:

$$\mathcal{E}(\gamma) = \int_{\gamma} ds, \tag{2.57}$$

which can be rewritten as:

$$\int_0^l \left(x'^2 + y'^2\right)^{\frac{1}{2}} ds, \tag{2.58}$$

Applying Eq. (2.37) using:

$$g(s, x, y, x', y') = \left(\left(x'(s)\right)^2 + \left(y'(s)\right)^2\right)^{\frac{1}{2}}, \tag{2.59}$$

where s is the arc length parameter, gives:

$$\begin{aligned} \frac{\partial \mathcal{E}}{\partial x} &= \frac{\partial g}{\partial x} - \frac{d}{ds}\left(\frac{\partial g}{\partial x'}\right) = -\frac{d}{ds}\left(\frac{x'}{\left((x')^2 + (y')^2\right)^{1/2}}\right) = -\frac{dx'}{ds} \\ \frac{\partial \mathcal{E}}{\partial y} &= \frac{\partial g}{\partial y} - \frac{d}{ds}\left(\frac{\partial g}{\partial y'}\right) = -\frac{d}{ds}\left(\frac{y'}{\left((x')^2 + (y')^2\right)^{1/2}}\right) = -\frac{dy'}{ds}, \end{aligned} \tag{2.60}$$

where we have used the fact that when a curve $\mathbf{c}$ is parametrized by arc length then $\|\mathbf{c}'\| = 1$. Equation (2.60) are written in vector form as

$$\frac{\partial \mathcal{E}}{\partial \gamma} = -\frac{d\mathbf{t}}{ds}, \tag{2.61}$$

where $\mathbf{t}$ is the unit tangent vector of γ. Using the definition Eq. (2.4) of curvature and assuming the configuration of Fig. 2.1 where the curve is oriented clockwise and the normal outward, we have:

$$\frac{\partial \mathcal{E}}{\partial \gamma} = -\kappa \mathbf{n}, \tag{2.62}$$

If we orient the curve in the opposite direction, i.e., counter clockwise, but leave the normal pointing outward, then:

$$\frac{\partial \mathcal{E}}{\partial \gamma} = \kappa \mathbf{n}, \tag{2.63}$$

where the curvature κ is still given by Eq. (2.7).

Path Integral of a Scalar Function (Two Dimensions)

Consider the following functional:

$$\mathscr{E}(\gamma) = \int_{\gamma} h ds, \tag{2.64}$$

where h is a scalar function, i.e., independent of γ. For $h = 1$ we have the curve length integral Eq. (2.57) as a special case. We can rewrite this functional as:

$$\mathscr{E}(\gamma) = \int_0^l h \left((x')^2 + (y')^2\right)^{1/2} ds, \tag{2.65}$$

Using

$$g(s, x, y, x', y') = h(x(s), y(s)) \left(\left(x'(s)\right)^2 + \left(y'(s)\right)^2\right)^{\frac{1}{2}}, \tag{2.66}$$

where s is the arc length parameter, the functional derivative of $\mathscr{E}$ with respect to the component x of γ is developed as follows:

$$\begin{aligned}\frac{\partial \mathscr{E}}{\partial x} &= \frac{\partial g}{\partial x} - \frac{d}{ds}\left(\frac{\partial g}{\partial x'}\right) \\ &= h_x - (\nabla h \cdot \mathbf{t})\, x' - h\frac{dx'}{ds},\end{aligned} \tag{2.67}$$

where $\mathbf{t}$ is the unit tangent vector of γ. Similarly, we have:

$$\frac{\partial \mathscr{E}}{\partial y} = h_y - (\nabla h \cdot \mathbf{t})\, y' - h\frac{dy'}{ds}$$

In vector form, we have:

$$\frac{\partial \mathscr{E}}{\partial \gamma} = \nabla h - (\nabla h \cdot \mathbf{t})\, \mathbf{t} - h\frac{d\mathbf{t}}{ds} \tag{2.68}$$

Since

$$\nabla h - (\nabla h \cdot \mathbf{t})\, \mathbf{t} = (\nabla h \cdot \mathbf{n})\, \mathbf{n} \tag{2.69}$$

and, according to the definition of curvature, $d\mathbf{t}/ds = \kappa \mathbf{n}$, we finally get:

$$\frac{\partial \mathscr{E}}{\partial \gamma} = (\nabla h \cdot \mathbf{n} - h\kappa)\, \mathbf{n}. \tag{2.70}$$

Here also the formula assumes the configuration of Fig. 2.1 where the curve is oriented clockwise and the normal outward. If we orient the curve in the opposite direction, i.e., counter clockwise, but leave the normal pointing outward, then:

$$\frac{\partial \mathscr{E}}{\partial \gamma} = (\nabla h \cdot \mathbf{n} + h\kappa)\,\mathbf{n}. \tag{2.71}$$

Next we give generic derivations of the functional derivatives of surface integrals of a scalar function and volume integrals of a scalar function [4]. Such integrals appear in the objective functional Eq. (4.80) in Chap. 5.

Surface Integral of a Scalar Function

We will now develop the functional derivative of a surface integral of a scalar function [4]:

$$\mathscr{E}_1(S) = \int_S g\, d\sigma, \tag{2.72}$$

where S is a closed regular surface in $\mathbb{R}^3$ and g is a scalar function independent of S. Let $(O, \mathbf{i}, \mathbf{j}, \mathbf{k})$ be a cartesian reference system in $\mathbb{R}^3$, and ϕ a parameterization of S:

$$\phi : (r, s) \in [0, l_1] \times [0, l_2] \to \phi(r, s) = (x(r, s), y(r, s), z(r, s)) \in \mathbb{R}^3 \tag{2.73}$$

Let $\mathbf{T}_r$ and $\mathbf{T}_s$ be the following vectors:

$$\begin{aligned} \mathbf{T}_r &= x_r\mathbf{i} + y_r\mathbf{j} + z_r\mathbf{k} \\ \mathbf{T}_s &= x_s\mathbf{i} + y_s\mathbf{j} + z_s\mathbf{k}, \end{aligned} \tag{2.74}$$

where the subscripts on x, y, z indicate partial derivatives. Functional Eq. (2.72) can be rewritten as [1, 3]:

$$\mathscr{E}_1(S) = \int_0^{l_1}\int_0^{l_2} g(x, y, z)\,\|\mathbf{T}_r \times \mathbf{T}_s\|\, drds, \tag{2.75}$$

Let L_1 designate the integrand of $\mathscr{E}_1$:

$$L_1(x, y, z, x_r, y_r, z_r, x_s, y_s, z_s, r, s) = g\,\|\mathbf{T}_r \times \mathbf{T}_s\| \tag{2.76}$$

The functional derivatives corresponding to $\mathscr{E}_1$ follow from the formulas:

$$\begin{aligned}
\frac{\partial \mathscr{E}_1}{\partial x} &= \frac{\partial L_1}{\partial x} - \frac{\partial}{\partial r}\frac{\partial L_1}{\partial x_r} - \frac{\partial}{\partial s}\frac{\partial L_1}{\partial x_s} \\
\frac{\partial \mathscr{E}_1}{\partial y} &= \frac{\partial L_1}{\partial y} - \frac{\partial}{\partial r}\frac{\partial L_1}{\partial y_r} - \frac{\partial}{\partial s}\frac{\partial L_1}{\partial y_s} \\
\frac{\partial \mathscr{E}_1}{\partial z} &= \frac{\partial L_1}{\partial z} - \frac{\partial}{\partial r}\frac{\partial L_1}{\partial z_r} - \frac{\partial}{\partial s}\frac{\partial L_1}{\partial z_s}.
\end{aligned} \tag{2.77}$$

Let $\mathbf{n}$ be the unit normal vector that points outwards to the exterior of S, i.e., toward the complement of its interior R_S, and let Φ be an orientation-preserving parametrization so that:

$$\mathbf{n} = \frac{\mathbf{N}}{\|\mathbf{N}\|} = \frac{\mathbf{T}_r \times \mathbf{T}_s}{\|\mathbf{T}_r \times \mathbf{T}_s\|}, \tag{2.78}$$

in which case:

$$L_1 = g\,\|\mathbf{T}_r \times \mathbf{T}_s\| = g\,\|\mathbf{N}\|\,. \tag{2.79}$$

Consider the formula of the first row of Eq. (2.77). We have the following developments:

$$\begin{aligned}
\frac{\partial L_1}{\partial x} &= g_x\,\|\mathbf{N}\| \\
\frac{\partial}{\partial r}\frac{\partial L_1}{\partial x_r} &= (\nabla g \cdot \mathbf{T}_r)\,\mathbf{n} \cdot \frac{\partial \mathbf{N}}{\partial x_r} + g\mathbf{n}_r \cdot \frac{\partial \mathbf{N}}{\partial x_r} \\
\frac{\partial}{\partial s}\frac{\partial L_1}{\partial x_s} &= (\nabla g \cdot \mathbf{T}_s)\,\mathbf{n} \cdot \frac{\partial \mathbf{N}}{\partial x_s} + g\mathbf{n}_s \cdot \frac{\partial \mathbf{N}}{\partial x_s}
\end{aligned} \tag{2.80}$$

The other two lines of Eq. (2.77) are developed in the same manner and we get:

$$\begin{aligned}
\frac{\partial \mathscr{E}_1}{\partial x} &= g_x\,\|\mathbf{N}\| - (\nabla g \cdot \mathbf{T}_r)\left(\mathbf{n} \cdot \frac{\partial \mathbf{N}}{\partial x_r}\right) - (\nabla g \cdot \mathbf{T}_s)\,\mathbf{n} \cdot \frac{\partial \mathbf{N}}{\partial x_s} - g\left(\frac{\partial \mathbf{N}}{\partial x_r} \cdot \mathbf{n}_r + \frac{\partial \mathbf{N}}{\partial x_s} \cdot \mathbf{n}_s\right) \\
\frac{\partial \mathscr{E}_1}{\partial y} &= g_y\,\|\mathbf{N}\| - (\nabla g \cdot \mathbf{T}_r)\left(\mathbf{n} \cdot \frac{\partial \mathbf{N}}{\partial y_r}\right) - (\nabla g \cdot \mathbf{T}_s)\,\mathbf{n} \cdot \frac{\partial \mathbf{N}}{\partial y_s} - g\left(\frac{\partial \mathbf{N}}{\partial y_r} \cdot \mathbf{n}_r + \frac{\partial \mathbf{N}}{\partial y_s} \cdot \mathbf{n}_s\right) \\
\frac{\partial \mathscr{E}_1}{\partial z} &= g_z\,\|\mathbf{N}\| - (\nabla g \cdot \mathbf{T}_r)\left(\mathbf{n} \cdot \frac{\partial \mathbf{N}}{\partial z_r}\right) - (\nabla g \cdot \mathbf{T}_s)\,\mathbf{n} \cdot \frac{\partial \mathbf{N}}{\partial z_s} - g\left(\frac{\partial \mathbf{N}}{\partial z_r} \cdot \mathbf{n}_r + \frac{\partial \mathbf{N}}{\partial z_s} \cdot \mathbf{n}_s\right)
\end{aligned} \tag{2.81}$$

Since $\mathbf{N} = \mathbf{T_r} \times \mathbf{T_s}$, we further have, looking back at the expression of $\mathbf{T}_r$ and $\mathbf{T}_s$ in Eq. (2.74):

$$\mathbf{n} \cdot \frac{\partial \mathbf{N}}{\partial x_r} = \mathbf{n} \cdot (\mathbf{i} \times \mathbf{T_s}) = (\mathbf{T_s} \times \mathbf{n}) \cdot \mathbf{i} \tag{2.82}$$

and, similarly, we have:

$$\mathbf{n} \cdot \frac{\partial \mathbf{N}}{\partial y_r} = (\mathbf{T}_s \times \mathbf{n}) \cdot \mathbf{j} \tag{2.83}$$

$$\mathbf{n} \cdot \frac{\partial \mathbf{N}}{\partial z_r} = (\mathbf{T}_s \times \mathbf{n}) \cdot \mathbf{k}, \tag{2.84}$$

giving the vectorial equation:

$$\begin{bmatrix} \mathbf{n} \cdot \frac{\partial \mathbf{N}}{\partial x_r} \\ \mathbf{n} \cdot \frac{\partial \mathbf{N}}{\partial y_r} \\ \mathbf{n} \cdot \frac{\partial \mathbf{N}}{\partial z_r} \end{bmatrix} = \mathbf{T}_s \times \mathbf{n} \tag{2.85}$$

Similar manipulations give:

$$\begin{bmatrix} \mathbf{n} \cdot \frac{\partial \mathbf{N}}{\partial x_s} \\ \mathbf{n} \cdot \frac{\partial \mathbf{N}}{\partial y_s} \\ \mathbf{n} \cdot \frac{\partial \mathbf{N}}{\partial z_s} \end{bmatrix} = \mathbf{n} \times \mathbf{T}_r, \tag{2.86}$$

and

$$\begin{bmatrix} \frac{\partial \mathbf{N}}{\partial x_r} \cdot \mathbf{n}_r \\ \frac{\partial \mathbf{N}}{\partial y_r} \cdot \mathbf{n}_r \\ \frac{\partial \mathbf{N}}{\partial z_r} \cdot \mathbf{n}_r \end{bmatrix} = \mathbf{T}_s \times \mathbf{n}_r; \begin{bmatrix} \frac{\partial \mathbf{N}}{\partial x_s} \cdot \mathbf{n}_s \\ \frac{\partial \mathbf{N}}{\partial y_s} \cdot \mathbf{n}_s \\ \frac{\partial \mathbf{N}}{\partial z_s} \cdot \mathbf{n}_s \end{bmatrix} = \mathbf{n}_s \times \mathbf{T}_r \tag{2.87}$$

We substitute Eqs. (2.85)–(2.87) in Eq. (2.81) to get the following vectorial expression of the functional derivative of $\mathscr{E}_1$:

$$\frac{\partial \mathscr{E}_1}{\partial \mathbf{x}} = \|\mathbf{N}\| \, \nabla g - (\nabla g \cdot \mathbf{T}_r)\,(\mathbf{T}_s \times \mathbf{n}) - (\nabla g \cdot \mathbf{T}_s)\,(\mathbf{n} \times \mathbf{T}_r) - g\,(\mathbf{T}_s \times \mathbf{n}_r + \mathbf{n}_s \times \mathbf{T}_r)\,, \tag{2.88}$$

where $\mathbf{x} = (x, y, z)$. This is not yet the expression we want and proceed to further developments. We decompose ∇g in the first term of the right-hand side of equation Eq. (2.88) in the basis $\left(\frac{\mathbf{T}_r}{\|\mathbf{T}_r\|}, \frac{\mathbf{T}_s}{\|\mathbf{T}_s\|}, \mathbf{n}\right)$, and we express $\mathbf{n}_r$ and $\mathbf{n}_s$ as a linear combination of $\mathbf{T}_r$ and $\mathbf{T}_s$ [1]:

$$\begin{aligned} \mathbf{n}_r &= a_{11}\mathbf{T}_r + a_{12}\mathbf{T}_s \\ \mathbf{n}_s &= a_{12}\mathbf{T}_r + a_{22}\mathbf{T}_s, \end{aligned} \tag{2.89}$$

which, by substitution in Eq. (2.88) gives:

$$\begin{aligned}\frac{\partial \mathscr{E}_1}{\partial \mathbf{x}} = &\ \|\mathbf{N}\| \left((\nabla g \cdot \mathbf{n})\, \mathbf{n} + (\nabla g \cdot \mathbf{t}_r)\, \mathbf{t}_r + (\nabla g \cdot \mathbf{t}_s)\, \mathbf{t}_s\right) \\ &- (\nabla g \cdot \mathbf{t}_r)\,(\mathbf{t}_s \times \mathbf{n})\, \|\mathbf{t}_s\|\, \|\mathbf{t}_r\| - (\nabla g \cdot \mathbf{t}_s)\,(\mathbf{n} \times \mathbf{t}_r)\, \|\mathbf{T}_s\|\, \|\mathbf{T}_r\| \\ &- g\,(a_{11} \mathbf{T}_s \times \mathbf{T}_r + a_{22} \mathbf{T}_s \times \mathbf{T}_r)\end{aligned} \tag{2.90}$$

Using circular permutations of the identity: $\mathbf{t}_r \times \mathbf{t}_s = \mathbf{n}$, and the definition of the mean curvature:

$$\kappa = \frac{1}{2}(a_{11} + a_{22}), \tag{2.91}$$

we finally get the desired expression of the functional derivative of the integral of a scalar function Eq. (2.72):

$$\frac{\partial \mathscr{E}_1}{\partial \mathbf{x}} = (\nabla g \cdot \mathbf{n} + 2g\kappa)\, \mathbf{N} \tag{2.92}$$

Volume Integral of a Scalar Function

We will now develop the functional derivative of a volume integral of a scalar function [4]:

$$\mathscr{E}_2\,(S) = \int_{V_S} f\, d\rho, \tag{2.93}$$

where V_S is the volume bounded by S. We will first transform $\mathscr{E}_2$ into a surface using the Gauss' divergence theorem. To do this, let $\mathbf{F} = P\mathbf{i} + Q\mathbf{j} + R\mathbf{k}$ be the vector field defined by:

$$\begin{aligned} P\,(x, y, z) &= \tfrac{1}{3} \int_0^x f\,(\lambda, y, z)\, d\lambda \\ Q\,(x, y, z) &= \tfrac{1}{3} \int_0^y f\,(x, \lambda, z)\, d\lambda \\ R\,(x, y, z) &= \tfrac{1}{3} \int_0^z f\,(x, y, \lambda)\, d\lambda \end{aligned} \tag{2.94}$$

Then:

$$\text{div}\mathbf{F} = P_x + Q_y + R_z = f, \tag{2.95}$$

where subscripts indicate partial derivation. Using the Gauss divergence theorem we have [3]:

$$\int_{R_S} f d\rho = \int_{V_S} \mathrm{div}\mathbf{F} d\rho = \int_S \mathbf{F} \cdot \mathbf{n} \, d\sigma$$
$$= \int_0^{l_1} \int_0^{l_2} \mathbf{F} \cdot \mathbf{n} \, \|\mathbf{T}_r \times \mathbf{T}_s\| \, drds, \tag{2.96}$$

where $\mathbf{n}$ is the outward unit normal to S. Designate by $L_2(r, s, x, y, z, x_r, y_r, z_r, x_s, y_s, z_s)$ the integrand of the last integral in Eq. (2.96). Just as with $\mathscr{E}_1$, the functional derivative of $\mathscr{E}_2$ with respect to $\mathbf{x} = (x, y, z)$ follows the formulas:

$$\begin{aligned}
\frac{\partial \mathscr{E}_2}{\partial x} &= \frac{\partial L_2}{\partial x} - \frac{\partial}{\partial r}\frac{\partial L_2}{\partial x_r} - \frac{\partial}{\partial s}\frac{\partial L_2}{\partial x_s} \\
\frac{\partial \mathscr{E}_2}{\partial y} &= \frac{\partial L_2}{\partial y} - \frac{\partial}{\partial r}\frac{\partial L_2}{\partial y_r} - \frac{\partial}{\partial s}\frac{\partial L_2}{\partial y_s} \\
\frac{\partial \mathscr{E}_2}{\partial z} &= \frac{\partial L_2}{\partial z} - \frac{\partial}{\partial r}\frac{\partial L_2}{\partial z_r} - \frac{\partial}{\partial s}\frac{\partial L_2}{\partial z_s}
\end{aligned} \tag{2.97}$$

Developing $\mathbf{N}$ as:

$$\mathbf{N} = \mathbf{T}_r \times \mathbf{T}_s = \begin{pmatrix} x_r \\ y_r \\ z_r \end{pmatrix} \times \begin{pmatrix} x_s \\ y_s \\ z_s \end{pmatrix} = \begin{pmatrix} y_r z_s - y_s z_r \\ -x_r z_s + x_s z_r \\ x_r y_s - x_s y_r \end{pmatrix} = \begin{pmatrix} N_1 \\ N_2 \\ N_3 \end{pmatrix}, \tag{2.98}$$

we find that:

$$\begin{aligned}
\frac{\partial L_2}{\partial x} &= P_x N_1 + Q_x N_2 + R_x N_2 \\
\frac{\partial}{\partial r}\frac{\partial L_2}{\partial x_r} &= -Q_x x_r z_s - Q_y y_r z_s - Q_z z_r z_s + R_x x_r y_s + R_y y_r y_s + R_z z_r y_s \\
\frac{\partial}{\partial s}\frac{\partial L_2}{\partial x_s} &= Q_x x_s z_r + Q_y y_s z_r + Q_z z_s z_r - R_x x_s y_r - R_y y_s y_r - R_z z_s y_r
\end{aligned} \tag{2.99}$$

Using Eq. (2.98), substitution of Eq. (2.99) back in Eq. (2.97) gives:

$$\frac{\partial \mathscr{E}_2}{\partial x} = P_x N_1 + Q_y N_1 + R_z N_1 = f N_1 \tag{2.100}$$

Similar developments yield:

$$\begin{aligned}
\frac{\partial \mathscr{E}_2}{\partial y} &= f N_2 \\
\frac{\partial \mathscr{E}_2}{\partial z} &= f N_3,
\end{aligned} \tag{2.101}$$

and, finally, this gives the desired result:

$$\frac{\partial \mathscr{E}_2}{\partial \mathbf{x}} = f\mathbf{N} \tag{2.102}$$

An objective functional we will study in Chap. 5 for motion tracking in the spatiotemporal domain [4, 5] has the form:

$$\mathscr{E}(S) = \int_{V_S} f \, d\rho + \int_S g \, d\sigma \tag{2.103}$$

According to the formulas Eqs. (2.92) and (2.102), the Euler-Lagrange equation corresponding to this functional is:

$$(f + \nabla g \cdot \mathbf{n} + 2g\kappa)\,\mathbf{n} = 0 \tag{2.104}$$

2.3 Differentiation Under the Integral Sign

In addition to having functions as arguments, the integrands of the objective functionals we will study in this book can also depend on a parameter. Their minimization with respect to the parameter uses the *differentiation under the integral sign*, in which the differentiation and integration operators are interchanged, i.e., the derivative of the integral with respect to the parameter is the integral of the integrand derivative with respect to the parameter. In its elementary calculus version, sufficient to us, the theorem of differentiation under the integral sign is as follows:

Let $J = [a, b]$ be a compact interval of $\mathbb{R}$ and A a compact subset of $\mathbb{R}^N$. Let $(\alpha, \mathbf{x}) \to f(\alpha, \mathbf{x})$ be a continuous real function on $J \times A$. If f has a continuous partial derivative $\frac{\partial f}{\partial \alpha}(\alpha, \mathbf{x})$ on $J \times A$, then the real function:

$$\mathscr{E}(\alpha) = \int_A f(\mathbf{x}, \alpha) d\mathbf{x} \tag{2.105}$$

is C^1 on J and:

$$\frac{d\mathscr{E}}{d\alpha}(\alpha) = \frac{d}{d\alpha}\left(\int_A f(\mathbf{x}, \alpha) d\mathbf{x}\right) = \int_A \frac{\partial f}{\partial \alpha}(\mathbf{x}, \alpha) d\mathbf{x} \tag{2.106}$$

2.4 Descent Methods for Unconstrained Optimization

Descent methods, sometimes also called greedy methods, for unconstrained optimization of an objective function with respect to an argument, are iterative methods which decrease the objective function at each iteration by incremental modification of the argument.

2.4.1 Real Functions

Let $f : \mathbf{x} \in \mathbb{R}^N \rightarrow f(\mathbf{x}) \in \mathbb{R}$ be a C^1 real function of which we want to determine an unconstrained local minimum, assuming such a minimum exists. To do so, consider $\mathbf{x}$ to be a C^1 function of (algorithmic) time, $\mathbf{x} : \tau \geq 0 \rightarrow \mathbf{x}(\tau) \in \mathbb{R}^N$, and let g be the composition of f and $\mathbf{x}$: $g(\tau) = f(\mathbf{x}(\tau))$. We have

$$\frac{dg}{d\tau} = \nabla f \cdot \frac{d\mathbf{x}}{d\tau} \tag{2.107}$$

Therefore, if we vary $\mathbf{x}$ from an initial position $\mathbf{x}_0$ according to the evolution equation:

$$\frac{d\mathbf{x}}{d\tau}(\tau) = -\alpha \nabla f(\mathbf{x}(\tau)), \quad \alpha \in \mathbb{R}^+, \tag{2.108}$$

then f will always decrease because:

$$\frac{df}{d\tau}(\mathbf{x}(\tau)) = \frac{dg}{d\tau}(\tau) = -\alpha \|\nabla f(\mathbf{x}(\tau))\|^2 \leq 0, \tag{2.109}$$

and will eventually reach a local minimum. More generally, if we vary $\mathbf{x}$ in direction $\mathbf{d}$ according to:

$$\begin{aligned} &\frac{d\mathbf{x}}{d\tau}(\tau) = -\alpha(\tau)\mathbf{d}(\mathbf{x}(\tau)) \\ &\mathbf{x}(0) = \mathbf{x}_0, \end{aligned} \tag{2.110}$$

where $\alpha(\tau) \in \mathbb{R}^+$ and $\nabla f \cdot \mathbf{d} > 0$, then f will vary according to

$$\frac{df}{d\tau}(\mathbf{x}(\tau)) = \frac{dg}{d\tau}(\tau) = -\alpha(\tau)\nabla f(\mathbf{x}(\tau)) \cdot \mathbf{d}(\mathbf{x}(\tau)) \leq 0 \tag{2.111}$$

Methods of unconstrained minimization based on Eq. (2.110) are called *descent* methods. When $\mathbf{d} = \nabla f$ is used it is the *gradient*, or *fastest*, descent. The scaling function α is often predetermined. For instance, $\alpha(\tau) =$ constant, or $\alpha(\tau) = 1/\tau$. In general, numerical descent methods are implemented as [6]:

1. $k = 0; \quad \mathbf{x}^0 = \mathbf{x}_0$
2. Repeat until convergence

$$\begin{aligned} \mathbf{d}^k &= \mathbf{d}(\mathbf{x}^k) \\ \alpha_k &= \arg\min_{\alpha \geq 0} f(\mathbf{x}^k - \alpha \mathbf{d}^k) \\ \mathbf{x}^{k+1} &= \mathbf{x}^k - \alpha_k \mathbf{d}^k \\ k &\leftarrow k + 1 \end{aligned}$$

Vectorial functions $F = (f_1, ..., f_n)^t$ are processed similarly by treating each component real function f_i as described above.

2.4.2 Integral Functionals

Consider the problem of minimizing functional Eq. (2.27):

$$\mathscr{E}(u) = \int_{x_1}^{x_2} g(x, u, u')dx,$$

where $u = u(x)$ and u' is the derivative of u with respect to x. To do so, let u vary in time, i.e., u is embedded in a one-parameter family of functions indexed by (algorithmic) time τ, and consider the time-dependent functional:

$$\mathscr{E}(u, \tau) = \int_{x_1}^{x_2} g(x, u(x, \tau), u'(x, \tau))dx. \tag{2.112}$$

The derivative of $\mathscr{E}$ with respect to the time parameter τ develops as:

$$\begin{aligned}
\frac{\partial \mathscr{E}}{\partial \tau} &= \int_{x_1}^{x_2} \left(\frac{\partial g}{\partial u} \frac{\partial u}{\partial \tau} + \frac{\partial g}{\partial u'} \frac{\partial u'}{\partial \tau} \right) dx \\
&= \int_{x_1}^{x_2} \left(\frac{\partial g}{\partial u} \frac{\partial u}{\partial \tau} + \frac{\partial g}{\partial u'} \frac{\partial}{\partial \tau} \left(\frac{\partial u}{\partial x} \right) \right) dx \\
&= \int_{x_1}^{x_2} \left(\frac{\partial g}{\partial u} \frac{\partial u}{\partial \tau} + \frac{\partial g}{\partial u'} \frac{\partial}{\partial x} \left(\frac{\partial u}{\partial \tau} \right) \right) dx
\end{aligned}$$

Integration by parts of the second term of the integrand yields:

$$\frac{\partial \mathscr{E}}{\partial \tau} = \left. \frac{\partial g}{\partial u'} \frac{\partial u}{\partial \tau} \right]_{x_1}^{x_2} + \int_{x_1}^{x_2} \left(\frac{\partial g}{\partial u} - \frac{\partial}{\partial x} \left(\frac{\partial g}{\partial u'} \right) \right) \frac{\partial u}{\partial \tau} dx \tag{2.113}$$

Assuming the endpoint conditions

$$\frac{\partial u}{\partial \tau}(x_1, \tau) = \frac{\partial u}{\partial \tau}(x_2, \tau) \;\; \forall \tau, \tag{2.114}$$

we finally obtain:

$$\frac{\partial \mathscr{E}}{\partial \tau} - \int_{x_1}^{x_2} \left(\frac{\partial g}{\partial u} - \frac{\partial}{\partial x} \left(\frac{\partial g}{\partial u'} \right) \right) \frac{\partial u}{\partial \tau} dx \tag{2.115}$$

Therefore, when u varies according to the evolution equation:

$$\frac{\partial u}{\partial \tau} = -\left(\frac{\partial g}{\partial u} - \frac{\partial}{\partial x}\left(\frac{\partial g}{\partial u'}\right)\right), \tag{2.116}$$

i.e.,

$$\frac{\partial u}{\partial \tau} = -\frac{\partial \mathscr{E}}{\partial u}, \tag{2.117}$$

it implies that:

$$\frac{\partial \mathscr{E}}{\partial \tau} = -\int_{x_1}^{x_2} \left(\frac{\partial g}{\partial u} - \frac{\partial}{\partial x}\left(\frac{\partial g}{\partial u'}\right)\right)^2 \leq 0 \tag{2.118}$$

Therefore, $\mathscr{E}$ continually decreases and, starting from an initial approximation $u(0) = u_0$, u will converge to a local minimum of $\mathscr{E}$, assuming such a minimum exists. Evolution Eq. (2.116) is the fastest descent equation to minimize functional Eq. (2.112). Functionals of several dependent variables are processed similarly.

Example: Let $I : \Omega \subset \mathbb{R}^2 \rightarrow L \subset \mathbb{R}^+$ be an image and $\gamma : s \in [0, 1] \rightarrow (x(s), y(s)) \in \mathbb{R}$ a closed regular plane curve. Let $R_1 = R_\gamma$ be the interior of γ and $R_2 = R_\gamma^c$ its complement. Consider minimizing the following functional [7]:

$$\mathscr{E}(\gamma, \mu_1, \mu_2) = \int_{R_1} (I - \mu_1)^2 \, dxdy + \int_{R_2} (I - \mu_2)^2 \, dxdy + \lambda \int_\gamma ds, \tag{2.119}$$

where λ is a real constant and μ_1, μ_2 are real parameters. This is an image segmentation functional the minimization of which will realize a piecewise constant two-region partition of the image. The first two terms are data terms which evaluate the deviation of the image from a constant representation by μ_1 in R_1 and μ_2 in its complement R_2, and the length integral is a regularization term to promote shorter, smoother curves γ. The minimization of $\mathscr{E}$ can be done by an iterative two-step greedy algorithm which repeats two consecutive steps until convergence, one step to minimize with respect to the parameters μ_1 and μ_2 with γ fixed, and the other to minimize with respect to γ with μ_1, μ_2 fixed, i.e., assumed independent of γ. Minimization with respect to the parameters, with γ given, is done by setting the derivative of $\mathscr{E}$ with respect to each parameter to zero. The derivatives are obtained by differentiation under the integral sign and we have:

$$\frac{\partial \mathscr{E}}{\partial \mu_i} = \frac{\partial}{\partial \mu_i} \int_{R_i} (I - \mu_i)^2 \, dxdy = \int_{R_i} \frac{\partial}{\partial \mu_i} \left((I - \mu_i)^2\right) dxdy = 0, \quad i = 1, 2. \tag{2.120}$$

This immediately gives μ_i to be the mean value of I in R_i:

$$\mu_i = \frac{\int_{R_i} I(x, y) \, dxdy}{\int_{R_i} dxdy}, \qquad i = 1, 2. \tag{2.121}$$

The minimization of $\mathscr{E}$ with respect to γ assuming μ_1, μ_2 fixed, independent of γ thereof, can be done by embedding γ in a one-parameter family of curves $\gamma : s, \tau \in [0, 1] \times \mathbb{R}^+ \rightarrow \gamma(s, \tau) = (x(s, \tau), y(s, \tau), \tau) \in \Omega \times \mathbb{R}^+$ indexed by algorithmic time τ and using the corresponding (Euler-Lagrange) descent equation:

$$\frac{\partial \gamma}{\partial \tau} = -\frac{\partial \mathscr{E}}{\partial \gamma}. \tag{2.122}$$

Note that each of the component functions of γ satisfies the endpoint conditions Eq. (2.114). Orienting γ counterclockwise and its unit normal $\mathbf{n}$ to point away from its interior, writing the data term of R_2 as $\int_{\Omega}(I-\mu_2)^2 dxdy - \int_{R_1}(I-\mu_2)^2 dxdy$, and using the basic formulas of Eq. (2.56) and Eq. (2.63), we obtain the partial differential equation governing the evolution of γ in (algorithmic) time:

$$\frac{\partial \gamma}{\partial \tau} = -\left((I - \mu_1)^2 - (I - \mu_2)^2 + \lambda\kappa\right)\mathbf{n}, \tag{2.123}$$

Instead of taking μ_1, μ_2 fixed when deriving the minimization equation with respect to γ, one can substitute their expression Eq. (2.121) in the functional and then derive the equation, thereby accounting for the dependence of the parameters on γ, and then opting for gradient descent. In this particular case of functional, however, the terms in the calculations due to the dependence on γ cancel out and one ends up with the same equation as when simply assuming the parameters fixed, i.e., independent of curve variable γ. This is a general result for a dependence of parameters on γ of the type in Eq. (2.121) [8].

In the computer vision literature a curve such as γ is called an *active curve* or *active contour*, and a partial differential equation such as Eq. (2.122) is referred to as its *evolution equation*. It moves in the direction of its normal at every one of its points and at the speed specified by the factor multiplying $\mathbf{n}$ in the evolution equation.

A direct implementation of the evolution equation which would discretize the curve and move each of its points *explicitly* would, in general, run into insurmountable numerical difficulties. The *level set* implementation, which we take up next, is an efficient way of realizing curve evolution without the numerical ills of the explicit implementation.

2.5 Level Sets

The *level set method* [9, 10] is for problems of moving interfaces, for curves, or surfaces in higher dimensions, that are moved by a differential equation which affects their shape. From a general point of view, it is, therefore, about optimizing the shape of curves and surfaces. In the problems we address, curves and surfaces are made to move so as to adhere to the boundary of desired regions in an image. For example, to detect moving objects in an image sequence, an active curve can be made to move so as to coincide with image boundaries of high image motion contrast, which is a characteristic of moving object contours, and in motion-based image segmentation a number of such curves can be made to evolve so as to adhere to the boundary of distinctly moving objects in space.

For a "nice" smooth curve which keeps its shape approximately during motion, it is natural to think of following a number of maker points on the curve, simply moving them from their current position for a time step and then interpolating the resulting particle positions. However, such a simple means of processing a moving curve will generally come to unresolvable numerical ills. There are several reasons for this. The most obvious is perhaps the fact that the evolution equation can cause a change in the topology of the curve. For instance, the curve can *split* into two pieces or more. By following the points explicitly, i.e., individually, there is no general way to detect when this splitting occurs, in which case the curve in its constituent parts cannot be recovered and further processing of its motion will be unstable and arbitrarily erroneous. Similar numerical instability problems will assuredly occur when distinct curves *merge* to form a single one.

Two less obvious but nevertheless serious, and common, difficulties with explicit following of marker points on a moving curve are *fans* and *shocks*. Consider, as illustrated in Fig. 2.2a, a curve forming a corner initially and moving "outward" in the direction of its normal at constant speed, say at unit speed. The particles on the horizontal side of the corner move straight up and those on the vertical side straight to the left, all transported a unit distance away from their initial position. Between these two sets of makers, a gap, or fan, has developed where there is no information about the shape of the curve because there are no particles to move in that place. The gap will widen with every move outward and the process of following the curve can break down quickly.

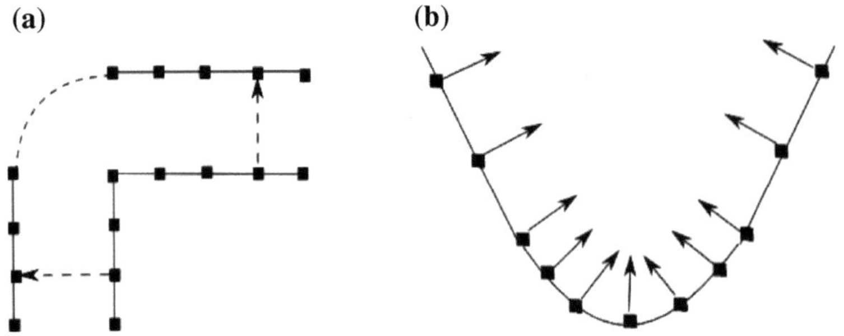

Fig. 2.2 Fanning and shocks cause instability when trying to follow the movement of curves via tracking explicitly a set of markers points on them. **a** Fanning: A corner moving outward in the direction of its normal at constant speed. The particles on the horizontal side of the corner move straight up and those on the vertical side straight to the left. Between these two sets of makers, a gap, or fan, has developed where we have no information about the actual shape of the curve. **b** shock: A curve with two straight segments, one on each side of a curved portion. When the marker points on the curved portion are moved inward, they are brought closer to each other and, with continued inward motion, will eventually meet and create a shock

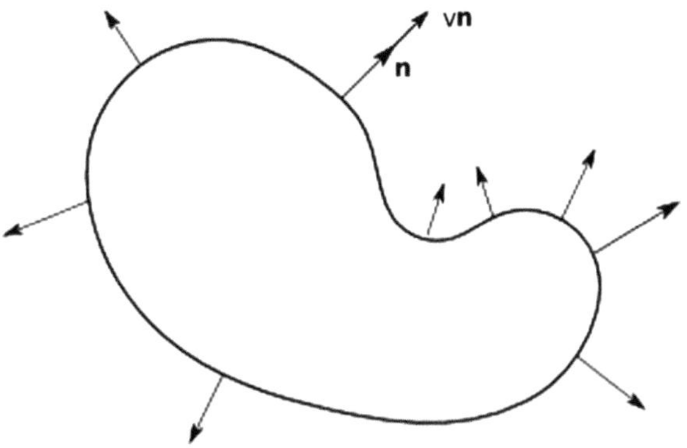

Fig. 2.3 A closed regular curve moving outward at speed V in the direction of its normal $\mathbf{n}$. Both $\mathbf{n}$ and V are functions of position on the curve

Shocks can appear when a contour initially curved as a fan is moved to "retract". This is illustrated in Fig. 2.2b, which basically shows the fanning example of Fig. 2.2a with time running in the opposite direction. The curve has two straight segments, one on each side of a curved portion. The particles in this curved portion will be brought closer to each other by inward motion and will eventually be so close as to meet and create a shock which will occult any previous ordering of the markers and, therefore, cause numerical griefs of various sorts.

The level set method, instead, moves active curves in a numerically stable manner. It deals efficiently with changes in the moving curve topology and with conditions such as fans and shocks when these occur. The basic idea is to describe the moving curve not explicitly by markers points on it but *implicitly* by a level set of a surface $\phi(x, y)$, the zero level set for instance, in which case the curve is represented for all practical purposes by $\phi(x, y) = 0$.

Let Γ be the set of closed regular plane curves $\gamma : s \in [0, 1] \rightarrow \gamma(s) \in \Omega$. For the purpose of describing its motion, an active contour is represented by a one-parameter family of curves in Γ indexed by algorithmic time t, i.e, a function $\gamma : s, \tau \in [0.1] \times \mathbb{R}^+ \rightarrow \gamma(s, \tau) = (x(s, \tau), y(s, \tau), \tau)) \in \Omega \times \mathbb{R}^+$ such that $\forall \tau$ curve $\gamma_\tau : s \rightarrow (x(s, \tau), y(s, \tau)$ is in Γ).

Consider a curve $\gamma \in \Gamma$ moving according to a velocity vector which is in the direction of its normal at each point (Fig. 2.3):

$$\frac{\partial \gamma}{\partial \tau} = V\mathbf{n}, \tag{2.124}$$

where V is the speed of motion. In the level set method, γ is described implicitly by the zero level set of a function $\phi : \mathbb{R}^2 \times \mathbb{R}^+ \rightarrow \mathbb{R}$ (Fig. 2.4):

$$\forall s, \tau \quad \phi(\gamma(s, \tau)) = \phi(x(s, \tau), y(s, \tau), \tau) = 0 \tag{2.125}$$

With ϕ sufficiently smooth, taking the total derivative of Eq. (2.125) with respect to time gives:

$$\frac{\partial \phi}{\partial \tau} = \nabla \phi \cdot \frac{\partial \gamma}{\partial \tau} + \frac{\partial \phi}{\partial \tau} = 0 \tag{2.126}$$

Using Eq. (2.124), we get:

$$\frac{\partial \phi}{\partial \tau} = -V \nabla \phi \cdot \mathbf{n} \tag{2.127}$$

With the convention that $\mathbf{n}$ is oriented outward and ϕ is positive inside its zero level set, negative outside, the normal $\mathbf{n}$ is given by:

$$\mathbf{n} = -\frac{\nabla \phi}{\|\nabla \phi\|}, \tag{2.128}$$

and substitution of Eq. (2.128) in Eq. (2.127) yields the evolution equation of ϕ:

$$\frac{\partial \phi}{\partial \tau} = V \|\nabla \phi\| \tag{2.129}$$

The analysis above applies to points on the level set zero of ϕ. Therefore, one must define *extension velocities* [10] to evolve the level set function elsewhere. Several possibilities have been envisaged. For instance, the extension velocity at a point has been taken to be the velocity of the evolving curve point closest to it. Extension velocities have also been defined so that the level set function is at all times the distance function from the evolving curve. In image segmentation and motion analysis problems, of the sort we have in this book, such extension velocities may not be easily implemented. When an expression of velocity is valid for all level sets, which is the case in just about all the active curve motion analysis methods in this book, then one can simply use this expression in all of the image domain, i.e., to evolve ϕ everywhere on its definition domain.

Regardless of what the extension velocities are chosen to be, the computational burden can be significantly lightened by restricting processing to a band around the active contour [10], a scheme called *narrow banding*.

By definition, an active curve γ can be recovered any time as the zero level set of its level set function ϕ and this is regardless of variations in its topology. The level set function always remains a function (Fig. 2.4), thereby assuring continued stable processing; stable processing is preserved also in the presence of fans and socks. In motion analysis problems which include motion-based image segmentation, and we will address some in this book, another advantage of the level set implementation is that region membership of points, i.e., the information as to which motion region they belong, is readily available from the sign of the level set functions.

There are several articles and books on the subject of level sets. The book of Sethian [10] is about efficient and numerically stable implementation of the level method, with examples from different domains of applications, including image analysis.

The velocities we will encounter in the forthcoming chapters of this book have components of one of three distinct types:

Type 1: V is a function of the curvature of the evolving curve. The component $\lambda\kappa$ in Eq. (2.123) is of this type.

Type 2: V is of the form $\mathbf{F} \cdot \mathbf{n}$ where $\mathbf{F}$ is a vector field dependent on position and possibly time but not on the curve. The term $\nabla h \cdot \mathbf{n}$ in Eq. (2.70), would it appear in a curve evolution equation, would be of this type. Such terms are called advection speeds in [10].

Type 3: V is a scalar function which depends on position and time but is not of the other two types. The velocity component $\left((I-\mu_1)^2-(I-\mu_2)^2\right)$ in Eq. (2.123), corresponding to the objective functional data terms, is of this type.

Velocities of the types 1, 2, and 3 are discretized differently as summarized below [10]. The velocity of a curve evolution in motion analysis is, in general, a linear combination of velocities of the three types. The discretization of Eq. (2.129) can be written in the following manner:

$$\phi_{ij}^{k+1} = \phi_{ij}^{k} + \Delta t \begin{cases} +V_{ij}^{k}\left((D_{ij}^{0x})^2 + (D_{ij}^{0y})^2\right)^{\frac{1}{2}} & \text{for type 1} \\ -\left(\max(F_{1ij}^{k},0)D_{ij}^{-x} + \min(F_{1ij}^{k},0)D_{ij}^{+x}\right. & \\ \left. + \max(F_{2ij}^{k},0)D_{ij}^{-y} + \min(F_{2ij}^{k},0)D_{ij}^{+y}\right) & \text{for type 2} \\ -\left(\max(V_{ij}^{k},0)\nabla^{+} + \min(V_{ij}^{k},0)\nabla^{-}\right) & \text{for type 3} \end{cases} \tag{2.130}$$

where i, j are indices on the discretization grid of Ω, k is the iteration index, F_1, F_2 are the coordinates of $\mathbf{F}$ appearing in the general expression of terms of type 2. Finite difference x−derivative operators D^{+x} (forward scheme), D^{-x} (backward scheme), and D^{0x} (central scheme), are applied to ϕ at i, j and iteration k, i.e., D_{ij}^{+x}, D_{ij}^{-x}, D_{ij}^{0x} in Eq. (2.130) stand for $D^{+x}(\phi^k)_{ij}$, $D^{-x}(\phi^k)_{ij}$, $D^{0x}(\phi^k)_{ij}$ and are given by:

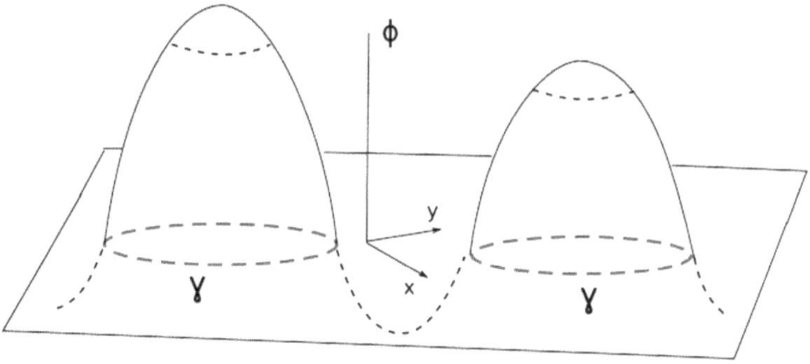

Fig. 2.4 The active curve γ is represented implicitly by the zero level of the level set function ϕ. Regardless of variations in the topology of γ, ϕ remains a function, thereby allowing stable curve evolution, unaffected by changes in the curve topology, fanning, and shocks. In this figure, γ has split into two component curves but ϕ remains a function

$$\begin{aligned} D_{ij}^{+x} &= \phi_{i+1,j}^k - \phi_{ij}^k \\ D_{ij}^{-x} &= \phi_{ij}^k - \phi_{i-1,j}^k \\ D_{ij}^{0x} &= \frac{1}{2}(\phi_{i+1,j}^k - \phi_{i-1,j}^k) \end{aligned} \tag{2.131}$$

Similar formulas and comments apply to the y−derivative operators D^{+y}, D^{-y}, and D^{0y}. Finally, operators ∇^+ and ∇^- are defined by:

$$\begin{aligned} \nabla^+ &= \Big(\max(D_{ij}^{-x}, 0)^2 + \min(D_{ij}^{+x}, 0)^2 \\ &\quad + \max(D_{ij}^{-y}, 0)^2 + \min(D_{ij}^{+y}, 0)^2\Big)^{\frac{1}{2}} \\ \nabla^- &= \Big(\max(D_{ij}^{+x}, 0)^2 + \min(D_{ij}^{-x}, 0)^2 \\ &\quad + \max(D_{ij}^{+y}, 0)^2 + \min(D_{ij}^{-y}, 0)^2\Big)^{\frac{1}{2}} \end{aligned} \tag{2.132}$$

The time step size Δt is adjusted for the experimentation at hand; it may vary for different applications. As a rule of thumb, one can chose its value so that the movement of the curve is approximately one pixel or less everywhere on the image positional array at each iteration.

References

1. M.P. Do Carmo, *Differential Geometry of Curves and Surfaces* (Prentice Hall, Englewood Cliffs, 1976)
2. R. Weinstock, *Calculus of Variations* (Dover, New York, 1974)
3. J.E. Marsden, A.J. Tromba, *Vector Calculus* (W. H. Freeman and Company, New York, 1976)
4. A. Mitiche, R. Feghali, A. Mansouri, Motion tracking as spatio-temporal motion boundary detection. J. Robot. Auton. Syst. **43**, 39–50 (2003)
5. R. El-Feghali, A. Mitiche, Spatiotemporal motion boundary detection and motion boundary velocity estimation for tracking moving objects with a moving camera: a level sets pdes approach with concurrent camera motion compensation. IEEE Trans. Image Process. **13**(11), 1473–1490 (2004)
6. M. Minoux, *Programmation mathématique*, vol. 1 (Dunod, Paris, 1983)
7. T. Chan, L. Vese, Active contours without edges. IEEE Trans. Image Process. **10**(2), 266–277 (2001)
8. G. Aubert, M. Barlaud, O. Faugeras, S. Jehan-Besson, Image segmentation using active contours: calculus of variations or shape gradients? SIAM J. Appl. Math. **63**(6), 2128–2154 (2003)
9. S. Osher, J. Sethian, Front propagation with curvature dependent speed: algorithms based on Hamilton-Jacobi formulations. J. Comput. Phys. **79**, 12–49 (1988)
10. J.A. Sethian, *Level set Methods and Fast Marching Methods* (Cambridge University Press, Cambridge, 1999)

Chapter 3
Optical Flow Estimation

3.1 Introduction

Optical flow is the velocity vector field of the projected environmental surfaces when a viewing system moves relative to the environment. Optical flow is a long standing subject of intensive investigation in diverse fields such as psychology, psychophysics, and computer vision [1–9]. In computer vision, of interest to us here, optical flow estimation has been a topic of continued interest and extensively researched. One of the most referenced paper on the subject is *Determining optical flow*, 1981, by B.K.P. Horn and B.G. Schunck [10]. It is also one of the most influential for having served as ground or benchmark for just about every dense flow computation algorithm. The Horn and Schunck variational formulation, which we will describe in detail subsequently (Sect. 3.4), seeks to determine the flow which minimizes a weighted sum of two integrals over the image domain, one to bring the flow to conform to the image spatiotemporal variations and the other to regularize the solution by constraining it to be smooth:

$$\mathscr{E}(u, v) = \int_{\Omega} (I_x u + I_y v + I_t)^2 dxdy + \lambda \int_{\Omega} (\|\nabla u\|^2 + \|\nabla v\|^2) dxdy, \qquad (3.1)$$

where $I : (x, y, t) \in \Omega \times]0, T[\mapsto I(x, y, t) \in \mathbf{R}^+$ is the image sequence of domain Ω and duration T, I_x, I_y, I_t its spatiotemporal derivatives, ∇u, ∇v the spatial gradients of the coordinates u, v of optical flow, and λ is a real constant to balance the contribution of the two terms in the functional. The corresponding Euler-Lagrange equations yield the flow via efficient implementation by Jacobi/Gauss-Seidel iterations.

A paper published the same year as [10] by B. D. Lucas and T. Kanade [11] on image registration and application to stereo-vision, has also been extensively referenced and used for optical flow estimation. The view taken was quite different as the scheme sought to determine the coordinate transformation between two images

A. Mitiche and J. K. Aggarwal, *Computer Vision Analysis of Image Motion by Variational Methods*, Springer Topics in Signal Processing 10, DOI: 10.1007/978-3-319-00711-3_3,

which minimized the displaced frame difference (DFD), i.e., the squared difference between one image and the other evaluated after the coordinate transformation (displaced, or warped as it is sometimes called). If the images are I_1 and I_2, and $\mathbf{x} \to \mathbf{f}(\mathbf{x}; \theta)$ is a parametric coordinate transformation with parameter vector θ, the scheme minimizes with respect to θ the objective function:

$$E(\theta) = \sum_{\mathbf{x} \in D} (I_1(\mathbf{f}(\mathbf{x}; \theta)) - I_2(\mathbf{x}))^2, \tag{3.2}$$

where D is a discretization of Ω. The minimization is carried out iteratively by expanding linearly I_1 at each step about the transformed coordinates of the previous step. The displacement at each point is computed subsequently from the estimated coordinate transformation: therefore, one of the significant conceptual differences between the methods of [10] and [11] is that the scheme in [10] references a vector field, i.e., a velocity vector as a variable at each point of the image domain, whereas the unknown in [11] is a global coordinate transformation between two images. Another difference is that the points at which there is no texture, i.e., where the spatial gradient is zero in the transformed image, do not contribute to determining the coordinate transformation whereas spatial regularization is a central concept in [10] which makes every point contribute to optical flow. From a computational point of view, the method of [11] involves a coordinate transformation and evaluation of the transformed image via spatial interpolation, an operation which does not occur in [10]. Often, the transformation has been applied locally in windows to allow spatial variations of the parameter vector estimate, which would improve its accuracy, but the window size affects the outcome which also suffers from the so-called block effect due to the lack of spatial regularization. However, both schemes [10, 11] have been combined in a continuous variational framework [12].

The Horn and Schunck algorithm solves a large but significantly sparse system of linear equations, which can be done very efficiently by convergent Jacobi or Gauss-Seidel iterations, particularly block-wise iterations [13]. A parallel hardware version has also been implemented [14, 15]. However, the basic neighborhood operations which drive the algorithm blur the optical flow estimate at motion boundaries. This serious problem is caused by the quadratic smoothness regularization term of the objective functional which leads to a Laplacian operator in the Euler-Lagrange equations. The discrete version of the operator reduces to averaging the estimate locally, which has the undesirable effect of blurring the computed flow at motion boundaries. Therefore, studies have subsequently considered using motion boundary preserving spatial regularizations. The problem has been addressed from four different perspectives: image driven smoothing, robust statistics, boundary length penalties, and nonlinear diffusion.

With image driven smoothing, the view is that motion edges coincide or tend to coincide with the image intensity edges, which would then justify that image motion smoothing be mediated by the image gradient [16–20]. However, this may also cause undue smoothing of motion because motion edges do not always occur at intensity edges, although image edges generally occur at motion edges.

Along the vein of robust statistics [21], motion discontinuity preservation is based on the notion of *outliers*. From this viewpoint, basically, the underlying interpretation is that the optical flow values over an image positional array satisfy the spatiotemporal data constraint and are smoothly varying everywhere except at motion boundaries where there are treated as outliers [22, 23]. This led to the use of robust functions such as the L^1 norm, the truncated quadratic [24], and the Lorentzian, in lieu of the quadratic to evaluate the objective functional terms. If ρ is a robust function, a discrete objective function one can seek to minimize is:

$$E(u,v) = \sum_{\mathbf{x} \in D} \left((I_x u + I_y v + I_t)^2(\mathbf{x}) + \lambda \sum_{\mathbf{y} \in \mathcal{N}_\mathbf{x}} (\rho(u(\mathbf{x}) - u(\mathbf{y})) + \rho(v(\mathbf{x}) - v(\mathbf{y}))) \right), \tag{3.3}$$

where $\mathcal{N}_\mathbf{x}$ is a set of neighbors of $\mathbf{x}$ (e.g., the 4-neighborhood). Slightly more general expressions can be adopted [22, 23, 25]. From this view of the problem, the effect of the robust function is to reduce the influence of the outliers on the estimation of optical flow and, therefore, provide a better definition of motion boundaries where the outliers are anticipated.

Another view of motion boundary preservation is to reference motion edges in the formulation and introduce a motion boundary length penalty term in the objective functional. This has been done via a line process in Markov random field (MRF) modelling [26]. Such a formulation, where edges are referenced by the MRF line process, has led to applications to optical flow estimation [27–29]. The objective function data term is still as in Eq. (3.3) and the regularization has the form:

$$\lambda \sum_{\mathbf{x} \in D} \sum_{\mathbf{y} \in \mathcal{N}_\mathbf{x}} \left(\alpha (1 - l_{\mathbf{x},\mathbf{y}}) \| W(\mathbf{x}) - W(\mathbf{y}) \|^2 + \beta l_{\mathbf{x},\mathbf{y}} \right), \tag{3.4}$$

where $W = (u, v)$, α and β are constants, and $l_{\mathbf{x},\mathbf{y}} \in \{0, 1\}$ is the binary variable of the MRF line process to represent the motion boundaries. Motion edges can also be referenced in a functional with a boundary length penalty term [30, 31] as in image segmentation [32, 33].

Finally, motion discontinuity preservation can be investigated from the perspective of nonlinear diffusion [34]. The rationale, basically, is that spatial regularization should be selective by allowing isotropic smoothing inside motion regions where optical flow is thought to be smooth, i.e., varies little spatially, and inhibit it across motion boundaries [35–45]. In particular, the L_1 norm regularization, also called total variation regularization and often abbreviated TV, which has been extensively investigated in inverse problems [46], notably image restoration [47], has been used for continuous variational optical flow estimation in several studies [39, 42–45, 48]. To simplify the rather elaborate TV minimization algorithm, and expedite the implementation thereof, the absolute value function in TV regularization, $g(z) = |z|$, is often replaced by a function of the sort $g(z) = \sqrt{z^2 + \varepsilon^2}$, for some small ε.

In the context of nonlinear diffusion optical flow estimation, the study in [37, 38] is singular in that it investigated regularization functions g in the functional:

$$\mathscr{E}(u, v) = \int_{\Omega} (I_x u + I_y v + I_t)^2 dxdy + \lambda \int_{\Omega} (g(\|\nabla u\|) + g(\|\nabla v\|))dxdy \quad (3.5)$$

by analyzing conditions which impose isotropic smoothing within motion regions and smoothing along motion boundaries but not across. The analysis leads to functions of the sort $g(s) = 2\sqrt{(1+s^2)} - 2$ (Aubert), $g(z) = log(1+s^2)$ (Perona-Malik [49, 50]), and others.

The optical flow constraint, on which most formulations of optical flow estimation rely, refers to the image sequence temporal derivative. In practice, image motion is often of large extent, typically causing displacements of several pixels between consecutive views. As a result, the image temporal derivative may not be approximated accurately to bear on motion estimation. In such a case, motion estimation has been addressed efficiently by multiresolution/multigrid processing [22, 51, 52]. Multiresolution and multigrid processing are "multilevel" computations which solve a system of equations on a given discretization grid by solving smaller similar systems on grids at coarser discretization.

Optical flow estimation has also been cast in a framework of simultaneous motion estimation and segmentation [31, 53–61], where the purpose is to divide the image into regions corresponding to distinct motions. Joint estimation and segmentation accounts for motion boundaries since those coincide with motion region boundaries. However, the emphasis in segmentation is not necessarily on accurate motion estimation because motion regions can be distinguished using simple motion models, the piecewise constant or affine models for instance, which do not necessarily describe the fine variations of motion that may be occurring.

Finally, it is worth mentioning that disparity estimation in binocular images resembles optical flow estimation and both problems can be cast in similar variational formulations. As a result, benefits can accrue from their joint estimation in stereoscopic image sequences [62, 63].

The purpose in this chapter forthcoming sections is to provide a digest of optical flow estimation by variational methods. We will not review the very large literature but rather describe a number of methods that would expose the fundamental concepts underlying image motion estimation by variational methods. The important ideas presented include (i) the basic formulation of image motion estimation as the minimization of a functional containing a data term and a regularization term; (ii) the use of optical flow smoothness in the regularization term; (iii) the notion of a motion boundary and definitions of it which would allow its preservation; (iv) the representation of motion by parametric functions; (v) the relationship between motion estimation and motion-based segmentation; and (vi) the concepts of mutiresolution and multigrid computation and their role in processing large-magnitude motion. Motion in stereoscopy will also be brought up. This reductive, concept-oriented description of image motion estimation will be enhanced by references to recent studies which build upon the basic formulations we will discuss and provide important computational details.

We will start the presentation with the optical flow constraint (Sect. 3.2) and immediately follow with the benchmark algorithms of Lucas-Kanade (Sect. 3.3) and of Horn and Schunck (Sect. 3.4). Motion boundary preservation will be treated next with the scheme of Deriche, Aubert, and Kornprobst [37] (Sect. 3.5), followed by image intensity based regularization [20] (Sect. 3.6) and the minimum description length (MDL) [31] (Sect. 3.7) formulations. Section 3.8 will describe parametric motion representation and computation. After a brief mention of variants of the smoothness and regularization terms (Sect. 3.9), the chapter will continue with a discussion of multiresolution/multigrid processing (Sect. 3.10) and a presentation of joint optical flow estimation and segmentation [54] (Sect. 3.11). Motion estimation in stereoscopy will be investigated in Sect. 3.12. The chapter does not provide an experimental evaluation or comparison of the methods but it gives examples of results. Evaluations of methods can be found in some of the cited papers.

3.2 The Optical Flow Constraint

Let the image sequence be a C^1 function $I : (x, y, t) \in \Omega \times]0, T[\mapsto I(x, y, t) \in \mathbf{R}^+$. Let $\mathbf{P}$ be a point on an environmental surface and $\mathbf{p}$ its image with coordinates $x(t), y(t)$ at instant t. As $\mathbf{P}$ moves in space, let the spatiotemporal trajectory of $\mathbf{p}$ have the parametric representation $t \to \mathbf{c}(t) = (x(t), y(t), t)$. Let h be the function $t \to h(t) = I \circ \mathbf{c}(t) = I(x(t), y(t), t)$, where $\circ$ indicates function composition. Function h is the image intensity along the motion trajectory of $\mathbf{p}$. If we assume that the intensity recorded from the environmental point $\mathbf{P}$ does not change as the surface it lies on moves, i.e., if h is constant, then we have the *optical flow constraint* (OFC) at $\mathbf{p}$:

$$\frac{dh}{dt} = \frac{\partial I}{\partial x}\frac{dx}{dt} + \frac{\partial I}{\partial y}\frac{dy}{dt} + \frac{\partial I}{\partial t} = 0 \tag{3.6}$$

or, using the usual subscript notation for the partial derivatives:

$$I_x u + I_y v + I_t = 0, \tag{3.7}$$

where $(u, v) = (\frac{dx}{dt}, \frac{dy}{dt})$ is called the *optical velocity* vector. The field over Ω of optical velocities is the *optical flow*.

The assumption that the recorded intensity is constant along motion trajectories is valid for translating Lambertian surfaces under constant uniform lighting. It is generally accepted that it is a good approximation for small velocity motions of non specular surfaces occurring over a short period of time. There have been a few attempts at determining constraints other than the invariance of image intensity along motion trajectories [64–67] but, by and large, the Horn and Schunck OFC (or discrete writings of it) has been the basic constraint in optical flow studies.

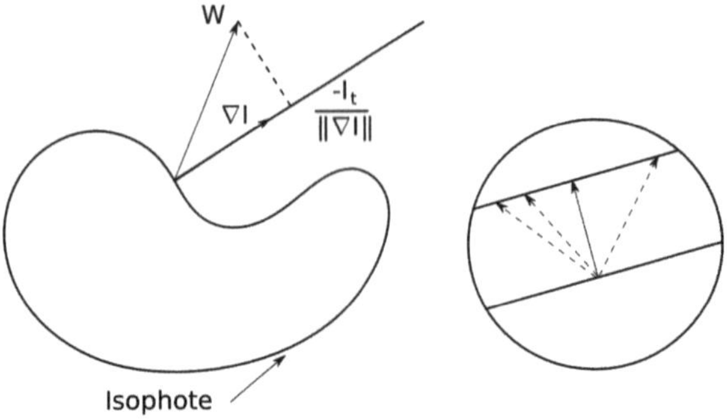

Fig. 3.1 *Left* The projection of the optical flow vector W on the image spatial gradient ∇I can be estimated from the image first-order spatiotemporal variations; whenever $\nabla I \neq 0$ it is equal to $-\frac{I_t}{\|\nabla I\|}$. *Right* The aperture problem: the movement of a straight edge seen through an aperture (a circular window in this figure) is ambiguous because only the component of motion in the direction perpendicular to the edge can be determined

If the optical velocity vector is denoted by W, the OFC is written $\nabla I \cdot W + I_t = 0$ and its projection $W^{\perp}$ in the direction of the image gradient, called the *normal component*, can be written:

$$W^{\perp} = \frac{\nabla I}{\|\nabla I\|} \cdot W = \frac{-I_t}{\|\nabla I\|}. \tag{3.8}$$

The spatiotemporal derivatives can be estimated from the image sequence data. Hence, the OFC determines the component of optical flow in the direction of the image gradient and only this component. This is a reflection of the *aperture problem*, the ambiguity in interpreting the translational motion of a straight line seen through an aperture, i.e., in the absence of any external visual cues (Fig. 3.1). The aperture problem is responsible for illusory percepts such as rotating spirals which appear to expand or contract and translating sine waves which appear highly non rigid [68]. The aperture problem was apprehended as early as 1911 by P. Stumpf [69] and has been the subject of many studies in psychophysics. In computer vision, it has been investigated in Hildreth's computational theory of visual motion measurement [70].

Local methods have been considered to solve the aperture problem. The simplest treatment assumes that optical flow is constant in the neighborhood of each point but that the image spatiotemporal gradient is not, leading to write one OFC for the same velocity at each point of the neighborhood [71]. Local processing of the aperture problem has also been addressed by the multiple OFC constraints method which assumes that there are $m \geq 2$ distinct image functions $I_1, ..., I_m$ satisfying the assumption of invariance to motion and giving m independent OFC equations to solve simultaneously. Several sources of these functions have been looked at [6]: (a) *multispectral images*, i.e., signals of different wavelengths as in colour images [72, 73], (b) *operators/filters*, where $I_1, ..., I_m$ are obtained by applying m operators/filters $O_1, ..., O_m$ to a single image function f. Examples include spatial filters applied to the original image [74] and differential operators [75–78]. Another source of functions is (c) *multiple illumination sources*, each giving a different image [79].

The assumptions supporting the local methods do not hold generally, leading to local systems of equations which are rank deficient or ill-conditioned. This is one important reason why variational methods which regularize the velocity field, such as those we reviewed in the introduction and some of which we will describe next, have been so much more effective.

3.3 The Lucas-Kanade Algorithm

The original study [11] addressed a general setting of image registration and developed an iterative algorithm which it applied to determining depth from stereoscopy. When used for optical flow evaluation, it has been applied in windows, typically 5×5 [80].

Let I_1 and I_2 be two images with the same domain Ω and $\mathbf{f} = (f_1, f_2)$ a coordinate transformation parametrized by $\theta = (\theta_1, ..., \theta_n) \in \mathbb{R}^n$:

$$\mathbf{f} : (\mathbf{x}, \theta) \in \Omega \times \mathbb{R} \rightarrow \mathbf{f}(\mathbf{x}, \theta) \in \Omega \tag{3.9}$$

Mapping $\mathbf{f}$ is often called a warp to distinguish it from a transformation that would act on the intensity image rather than on the image domain. The problem is to determine the transformation that minimizes the smallest displaced frame difference, i.e., determine $\tilde{\theta}$ such that:

$$\tilde{\theta} = \arg\min_{\theta} \sum_{\mathbf{x} \in D} (I_1(\mathbf{f}(\mathbf{x}; \theta)) - I_2(\mathbf{x}))^2 \tag{3.10}$$

The algorithm is developed from a first-order Taylor expansion of $I_1(\mathbf{f}(\mathbf{x}, \theta))$ with respect to θ. In a open neighborhood V of $\theta_0 \in \mathbb{R}$ we have, assuming $I_1(\mathbf{f}(\mathbf{x}, \theta))$ is differentiable in $\Omega \times V$,

$$I_1(\mathbf{f}(\mathbf{x}, \theta)) = I_1(\mathbf{f}(\mathbf{x}, \theta_0)) + \nabla I_1(\mathbf{x}, \theta_0) J_{\mathbf{f}}(\mathbf{x}, \theta_0)\mathbf{h} + o(\|\mathbf{h}\|^2), \tag{3.11}$$

where $\mathbf{h} = \theta - \theta_0$, ∇I_1 is the spatial gradient of I_1 written as a row vector, and $J_{\mathbf{f}}$ is the Jacobian of $\mathbf{f}$ with respect to θ:

$$J_{\mathbf{f}} = \begin{pmatrix} \frac{\partial f_1}{\partial \theta_1} & \cdots & \frac{\partial f_1}{\partial \theta_n} \\ \frac{\partial f_2}{\partial \theta_1} & \cdots & \frac{\partial f_2}{\partial \theta_n} \end{pmatrix} \tag{3.12}$$

Dropping the little o remainder, the objective function to minimize following this expansion is:

$$E(\mathbf{x},\theta) = \sum_{\mathbf{x}\in D} (I_1(\mathbf{f}(\mathbf{x},\theta_0)) + \nabla I_1(\mathbf{x},\theta_0) J_{\mathbf{f}}(\mathbf{x},\theta_0)\mathbf{h} - I_2(\mathbf{x}))^2 \tag{3.13}$$

Therefore, the first-order Taylor expansion has done two things: (1) the objective function turns into a linear equation in $\mathbf{h}$ and, (2) viewing θ_0 as a current estimate, its minimization turns into iterations which consist at each step of determining an update $\mathbf{h}$ by solving a linear system of equation by least squares. The scheme involves "warping" the image, i.e., evaluating I_1 at the points of grid D transformed by $\mathbf{f}$. In general, this involves interpolating the image I_1. The original paper [11] mentions solving for $\mathbf{h}$ using the least squares solution analytic expression, which involves matrix inversion, but other numerical schemes are more efficient, for instance the singular value decomposition method [81], and others which were investigated in the context of the Lucas-Kanade image registration algorithm [80]. The main weakness of the Lucas-Kanade formulation is its lack of regularization. In the Horn and Schunck formulation, which we review next, regularization of the flow field is fundamental.

3.4 The Horn and Schunck Algorithm

We recall the Horn and Schunck optical flow estimation functional [10] for an image sequence $I : (x, y, t) \in \Omega\times]0, T[\mapsto I(x, y, t) \in \mathbf{R}^+$:

$$\mathscr{E}(u, v) = \int_\Omega (I_x u + I_y v + I_t)^2 dxdy + \lambda \int_\Omega (\|\nabla u\|^2 + \|\nabla v\|^2) dxdy,$$

where I_x, I_y, I_t are the image spatiotemporal derivatives, $\nabla u, \nabla v$ are the spatial gradients of the optical flow coordinates u, v, and λ is a constant factor to weigh the contribution of the two terms in the objective functional. The corresponding Euler-Lagrange equations are:

$$\begin{aligned} I_x(I_x u + I_y v + I_t) - \lambda\nabla^2 u = 0 \\ I_y(I_x u + I_y v + I_t) - \lambda\nabla^2 v = 0, \end{aligned} \tag{3.14}$$

with Neumann boundary conditions

$$\frac{\partial u}{\partial \mathbf{n}} = 0, \quad \frac{\partial v}{\partial \mathbf{n}} = 0, \tag{3.15}$$

where $\frac{\partial}{\partial \mathbf{n}}$ designates differentiation in the direction of the normal $\mathbf{n}$ to the boundary of the image domain Ω, and ∇^2 denotes the Laplacian operator.

3.4.1 Discretization

Let D be a unit-spacing grid over Ω with the grid points indexed left-to-right and top-down by the integers $\{1, 2, ..., N\}$. For all grid point indices $i \in \{1, 2, ..., N\}$, a discrete approximation of the Euler-Lagrange Equations Eq. (3.14) is :

$$\begin{aligned} I_{xi}^2 u_i + I_{xi} I_{yi} v_i + I_{xi} I_{ti} - \lambda \sum_{j \in \mathcal{N}_i} (u_j - u_i) = 0 \\ I_{yi} I_{xi} u_i + I_{yi}^2 v_i + I_{yi} I_{ti} - \lambda \sum_{j \in \mathcal{N}_i} (v_j - v_i) = 0, \end{aligned} \tag{3.16}$$

where λ has absorbed the averaging constant of the Laplacian approximation; $(u_i, v_i) = (u, v)_i$ is the optical flow vector at grid point i; I_{xi}, I_{yi}, I_{ti} are the spatiotemporal derivatives I_x, I_y, I_t evaluated at i; and $\mathcal{N}_i$ is the set of indices of the neighbors of i for some neighborhood system. For the 4-neighborhood, for instance, $card(\mathcal{N}_i) < 4$ for pixels on the boundary of D and $card(\mathcal{N}_i) = 4$ for interior pixels. By accounting for the cardinality of $\mathcal{N}_i$, the approximation of the Laplacian in Eq. (3.16) is consistent with the Neumann boundary conditions Eq. (3.15) because it is equivalent to considering neighbors j of i outside the image domain but giving these the same flow vector as i. This is sometime called mirroring.

Re-arranging terms in Eq. (3.16), we have the following linear system of equations, for $i \in \{1, ..., N\}$:

$$(S) \begin{cases} (I_{xi}^2 + \lambda c_i) u_i + I_{xi} I_{yi} v_i - \lambda \sum_{j \in \mathcal{N}_i} u_j = -I_{xi} I_{ti} \\ I_{xi} I_{yi} u_i + (I_{yi}^2 + \lambda c_i) v_i - \lambda \sum_{j \in \mathcal{N}_i} v_j = -I_{yi} I_{ti}, \end{cases}$$

where $c_i = card(\mathcal{N}_i)$. Let $\mathbf{z} = (z_1, ..., z_{2N})^t \in \mathbf{R}^{2N}$ be the vector defined by

$$z_{2i-1} = u_i, \quad z_{2i} = v_i, \quad i \in \{1, ..., N\}. \tag{3.17}$$

Also, let $\mathbf{b} = (b_1, ..., b_{2N})^t \in \mathbf{R}^{2N}$ be defined by

$$b_{2i-1} = -I_{xi} I_{ti}, \quad b_{2i} = -I_{yi} I_{ti}, \quad i \in \{1, ..., N\}. \tag{3.18}$$

In matrix form, linear system (S) is:

$$\mathbf{Az} = \mathbf{b}, \tag{3.19}$$

where $\mathbf{A}$ is the $2N \times 2N$ matrix the elements of which are, for $i \in \{1, ..., N\}$:

$$\begin{aligned} \mathbf{A}_{2i-1,2i-1} = I_{xi}^2 + \lambda c_i, \quad \mathbf{A}_{2i,2i} = I_{yi}^2 + \lambda c_i, \\ \mathbf{A}_{2i-1,2i} = I_{xi} I_{yi}, \quad \mathbf{A}_{2i,2i-1} = I_{xi} I_{yi}, \\ \mathbf{A}_{2i-1,2j-1} = \mathbf{A}_{2i,2j} = -\lambda, \quad j \in \mathcal{N}_i, \end{aligned} \tag{3.20}$$

all other elements being equal to zero. System Eq. (3.19) is a large scale sparse system of linear equations. Such systems are best solved by iterative algorithms such as the Jacobi and Gauss-Seidel iterations [82, 83] which we will give next. We will assume that **A** is non-singular.

3.4.2 Gauss-Seidel and Jacobi Iterations

One can show that matrix **A** is positive definite [13]. This implies that the point-wise and block-wise Gauss-Seidel iterations to solve system Eq. (3.19) will converge. This is a standard result in numerical linear algebra [82, 83]. For the 2×2 block division of matrix **A**, the Gauss-Seidel iterations are [13], for all $i \in \{1, ..., N\}$:

$$\begin{aligned} u_i^{k+1} &= \frac{I_{yi}^2 + \lambda c_i}{c_i(I_{xi}^2 + I_{yi}^2) + \lambda c_i^2}\left(\sum_{j\in\mathcal{N}_i; j<i} u_j^{k+1} + \sum_{j\in\mathcal{N}_i; j>i} u_j^k\right) \\ &\quad - \frac{I_{xi}I_{yi}}{c_i(I_{xi}^2 + I_{yi}^2) + \alpha c_i^2}\left(\sum_{j\in\mathcal{N}_i; j<i} v_j^{k+1} + \sum_{j\in\mathcal{N}_i; j>i} v_j^k\right) - \frac{I_{xi}I_{ti}}{I_{xi}^2 + I_{yi}^2 + \lambda c_i} \end{aligned} \tag{3.21}$$

$$\begin{aligned} v_i^{k+1} &= \frac{-I_{xi}I_{yi}}{c_i(I_{xi}^2 + I_{yi}^2) + \lambda c_i^2}\left(\sum_{j\in\mathcal{N}_i; j<i} u_j^{k+1} + \sum_{j\in\mathcal{N}_i; j>i} u_j^k\right) \\ &\quad + \frac{I_{xi}^2 + \lambda c_i}{c_i(I_{xi}^2 + I_{yi}^2) + \lambda c_i^2}\left(\sum_{j\in\mathcal{N}_i; j<i} v_j^{k+1} + \sum_{j\in\mathcal{N}_i; j>i} v_j^k\right) - \frac{I_{yi}I_{ti}}{I_{xi}^2 + I_{yi}^2 + \lambda c_i} \end{aligned}$$

Horn and Schunck [10] solve system Eq. (3.19) with the 2×2 block-wise Jacobi method. The iterations are:

$$\begin{aligned} u_i^{k+1} &= \frac{I_{yi}^2 + \lambda c_i}{c_i(I_{xi}^2 + I_{yi}^2) + \lambda c_i^2}\sum_{j\in\mathcal{N}_i} u_j^k - \frac{I_{xi}I_{yi}}{c_i(I_{xi}^2 + I_{yi}^2) + \lambda c_i^2}\sum_{j\in\mathcal{N}_i} v_j^k - \frac{I_{xi}I_{ti}}{I_{xi}^2 + I_{yi}^2 + \lambda c_i} \\ v_i^{k+1} &= \frac{-I_{xi}I_{yi}}{c_i(I_{xi}^2 + I_{yi}^2) + \alpha c_i^2}\sum_{j\in\mathcal{N}_i} u_j^k + \frac{I_{xi}^2 + \lambda c_i}{c_i(I_{xi}^2 + I_{yi}^2) + \lambda c_i^2}\sum_{j\in\mathcal{N}_i} v_j^k - \frac{I_{yi}I_{ti}}{I_{xi}^2 + I_{yi}^2 + \lambda c_i} \end{aligned} \tag{3.22}$$

The fact that matrix **A** is symmetric positive definite is not sufficient to imply that the Jacobi iterations converge. However, it can be shown directly that they do. This has been done using a vector norm in $\mathbb{R}^{2N}$ adapted to the special structure of the linear system (S) [13].

The differences between the Gauss-Seidel and the Jacobi methods are well known: the Jacobi method does the update for all points of the image domain grid and then uses the updated values at the next iteration, whereas the Gauss-Seidel method uses

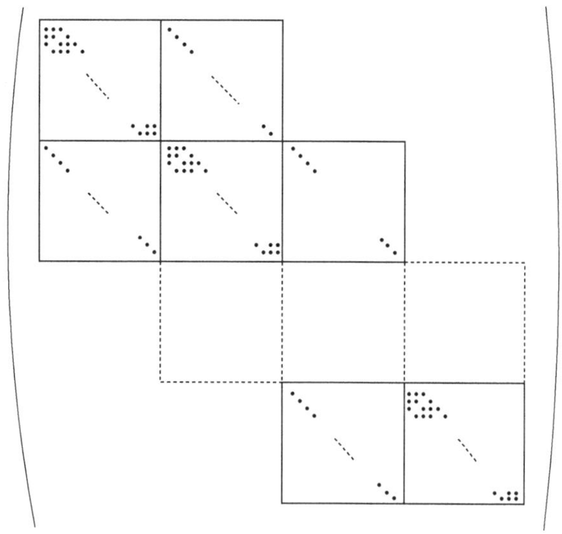

Fig. 3.2 Data matrix **A** is block tridiagonal. The dots represent possibly nonzero elements. For an $n \times n$ discrete image, the blocks are $2n \times 2n$. The block tridiagonal form comes from the fact that points with index $Kn, 1 \leq K \leq n$, do not have a right-side neighbor, and those with index $Kn + 1$, $0 \leq K \leq n - 1$, do not have a left-side neighbor

the updated values as soon as they are available and, as a result, can be more efficient than the Jacobi method in sequential computations. However, in contrast with the Gauss-Seidel iterations, Jacobi's can be performed in parallel for all pixels, which can result in a very fast hardware implementation [14, 15]. As to memory storage, the Jacobi method requires at each iteration the $2N$ values of the previous iteration in memory store. With the Gauss-Seidel iterations Eq. (3.22), only a few of these values are stored.

There is a remarkable block division which makes matrix **A** block tridiagonal (Fig. 3.2). Combined with the property that **A** is symmetric positive definite, this characteristic affords efficient resolution of the corresponding linear system [82]. For an $n \times n$ discrete image, the blocks are $2n \times 2n$. The block tridiagonal form is due to the fact that points with index Kn, $1 \leq K \leq n$, do not have a neighbor on the right, and those with index $Kn+1, 0 \leq K \leq n-1$, do not have a neighbor on the left. The block-wise iterations for a block tridiagonal symmetric positive definite matrix, i.e., the iterations corresponding to the tridiagonal block decomposition (Fig. 3.2), converge for both the Jacobi and the Gauss-Seidel implementations [82]. The spectral radius of the Gauss-Seidel matrix is equal to the square of the spectral radius of the Jacobi matrix, which signifies that the Gauss-Seidel implementation is in this case much faster that the Jacobi. The readers interested in the details may refer to [13].

3.4.3 Evaluation of Derivatives

Horn and Schunck have used approximations of the image spatial and temporal derivatives as averages of forward first differences. From two consecutive $n \times n$ images I^1 and I^2 the formulas are:

$$
\begin{aligned}
I_{xi} &= \frac{1}{4}(I^1_{i+1} - I^1_i + I^1_{i-n+1} - I^1_{i-n} + I^2_{i+1} - I^2_i + I^2_{i-n+1} - I^2_{i-n}) \\
I_{yi} &= \frac{1}{4}(I^1_{i-n} - I^1_i + I^1_{i-n+1} - I^1_{i+1} + I^2_{i-n} - I^2_i + I^2_{i-n+1} - I^2_{i+1}) \\
I_{ti} &= \frac{1}{4}(I^2_i - I^1_i + I^2_{i+1} - I^1_{i+1} + I^2_{i-n} - I^1_{i-n} + I^2_{i-n+1} - I^1_{i-n+1}),
\end{aligned}
\tag{3.23}
$$

for $i = 1, ..., n^2$. Alternatively, the spatial derivatives can be estimated using central differences. Using central differences to compute the temporal derivatives would not be consistent with the in-between consecutive frames velocities to be estimated because it would require using the frames preceding and following the current, rather than consecutive frames.

Points in the formulas which fall outside the image domain are often given the index of the image wrapped around on its boundary to form a (digital) torus or, alternatively, boundary points are simply given the spatiotemporal derivative values of an immediate interior neighbor.

3.4.4 Ad hoc Variations to Preserve Motion Boundaries

As alluded to in the introduction, the single serious drawback of the Horn and Schunck method is that the quadratic (Tikhonov) regularization it uses ignores motion boundaries which it smooths out as a result. This technically translates into the occurrence of the isotropic Laplacian operator in the Euler-Lagrange equations. The original study of Horn and Schunck approximates the discrete Laplacian $\Delta^2 w$ as:

$$
\Delta^2 w \propto \overline{w} - w, \tag{3.24}
$$

where w stands for either u or v and $\overline{w}$ is a weighted neighborhood average of w according to the weights in Fig. 3.3.

This Laplacian approximation is used explicitly in their discretization of the Euler-Lagrange equations to arrive at the following form of the iterations to compute optical flow, where λ has absorbed the coefficient of proportionality:

$$
\begin{aligned}
u_i^{k+1} &= \overline{u}_i{}^k - I_{xi}\frac{I_{xi}\overline{u}_i^k + I_{yi}\overline{v}_i^k + I_t}{\lambda + I_{xi}^2 + I_{yi}^2} \\
v_i^{k+1} &= \overline{v}_i{}^k - I_{yi}\frac{I_{xi}\overline{u}_i^k + I_{yi}\overline{v}_i^k + I_t}{\lambda + I_{xi}^2 + I_{yi}^2}
\end{aligned}
\tag{3.25}
$$

Boundary conditions aside, the average $\overline{w}$ is computed according to the set of fixed weights in Fig. 3.3. This suggests that one can be more general and approximate the operator by spatially variant filters, rather than a fixed weighted average, with the purpose of preserving motion boundaries, i.e., dampening blurring at motion discontinuities. In such a case, iterations Eq. (3.25) are executed with:

$$\begin{aligned} \overline{u}_i^k &= g\left(\{u_j^k\} : j \in \mathcal{N}_i\right) \\ \overline{v}_i^k &= g\left(\{v_j^k\} : j \in \mathcal{N}_i\right), \end{aligned} \tag{3.26}$$

with filters g such as those suggested in [84]. These can be dependent of the image or on the flow itself:

Image-based adaptive average: Under the assumption that the image of environmental objects is smooth except at the projection of their occluding boundaries, flow edges and image edges will coincide, justifying an intensity-based filter of the form:

$$g\left(\{w_j^k\} : j \in \mathcal{N}_i\right) = \sum_{j \in \mathcal{N}_i} \alpha_j w_j, \tag{3.27}$$

where coefficients α_j are commensurate with the image contrast between i and j, for instance by using:

$$\alpha_j = \frac{\frac{1}{1+|I_j - I_i|}}{\sum_{j \in \mathcal{N}_i} \frac{1}{1+|I_j - I_i|}} \tag{3.28}$$

In general, of course, flow discontinuities are only a subset of intensity edges so that smoothing of the flow field according to Eq. (3.28) will follow the image intensity structure rather than the structure of the motion field and, as a result, can cause undesirable artefacts.

Optical flow-based adaptive average: The coefficients of a flow-based version of the image-based filter would be:

$\frac{1}{12}$	$\frac{1}{6}$	$\frac{1}{12}$
$\frac{1}{6}$	-1	$\frac{1}{6}$
$\frac{1}{12}$	$\frac{1}{6}$	$\frac{1}{12}$

Fig. 3.3 The discrete Laplacian $\Delta^2 w$ can be written as $\Delta^2 w \propto \overline{w} - w$, where w stands for either u or v and $\overline{w}$ is a weighted neighborhood average of w using the weights above as suggested in the original investigation of optical flow estimation by the Horn and Schunck method

$$\alpha_j = \frac{\left(\frac{1}{1+|w_j - w_i|}\right)^\beta}{\sum_{j \in \mathcal{N}_i} \left(\frac{1}{1+|w_j - w_i|}\right)^\beta} \tag{3.29}$$

where w stands for either of the optical flow coordinates and $\beta > 1$. The purpose of exponent β is to discern better the coefficients values when the range of the flow coordinates is small. This filter is expected to dampen smoothing across motions discontinuities while stressing it along.

Median filtering: Here, filter g at pixel i would be the median of the current flow estimates in the neighborhood $\mathcal{N}_i$ of i. At a flow discontinuity, median filtering is more likely to yield a value representative of the values on a single side of the discontinuity. A reasonable alternative consists of averaging the values of the flow velocity in $\mathcal{N}_i$ which are above or below the median, whichever are more homogeneous. In the event of a flow edge at i, these values would most likely come from pixels on a single side of the edge.

Modulating the weight coefficient λ: The ad hoc variations above use digital approximations of the Laplacian which adjust to the local structure of the image or of the flow field at any stage of its estimation by the Horn and Schunck iterations, in the hope that this structure is actually indicative of the actual flow discontinuities. Along this vein of thought, one can also look at varying the weighing coefficient λ during the iterations depending on the structure of the image or the current flow field [85]. Since smoothing increases with λ, the rationale is that the value of this coefficient should be low at suspected motion boundaries and high elsewhere. For instance the study in [85] uses thresholds on $\|\nabla I\|^2$ and $\|\nabla u\|^2 + \|\nabla v\|^2$ to decide whether to smooth sufficiently, according to some threshold λ_h, when neither of these gradient norms is high or, instead, inhibit smoothing using a small coefficient λ_s.

Although ad hoc approximations of key variables in Eq. (3.25), such as the Laplacian or the weight coefficient, can produce sharper motion boundaries at practically no additional computational expense, there have been no extensive experimental verification which would allow a definite conclusion about their effectiveness compared to other boundary preserving formulations such as the ones we will describe next. These are formal methods which aim at preserving motion discontinuities by referencing motion edges via boundary length or by using a boundary preserving regularization function in the objective functional. We will describe both an image-based and a flow-based boundary preserving regularization function.

3.5 Deriche–Aubert–Kornprobst Method

The Laplacian operator which appears in the Euler-Lagrange equations associated with Eq. (3.14) causes smoothing, and blurring thereof, across motion boundaries. To circumvent the problem, the study in [37] proposed to investigate regularization functions g in the following generalization of the Horn and Schunck functional:

$$E(u, v) = \int_{\Omega} (I_x u + I_y v + I_t)^2 dxdy + \lambda \int_{\Omega} (g(\|\nabla u\|) + g(\|\nabla v\|))dxdy, \quad (3.30)$$

such that motion boundaries are preserved. With $g(z) = z^2$, Eq. (3.30) reduces to the Horn and Schunck functional Eq. (3.14). The purpose of the analysis in [37, 38] was to determine g from conditions that would ensure isotropic smoothing of motion where it varies smoothly and allow smoothing along motion boundaries while inhibiting or dampening it across. The analysis is summarized in the following.

The Euler-Lagrange equations corresponding to Eq. (3.30) are:

$$\begin{aligned} I_x(I_x u + I_y v + I_t) - \frac{\lambda}{2}\mathrm{div}\left(g'(\|\nabla u\|)\frac{\nabla u}{\|\nabla u\|}\right) &= 0 \\ I_y(I_x u + I_y v + I_t) - \frac{\lambda}{2}\mathrm{div}\left(g'(|\ \nabla v\|)\frac{\nabla v}{|\ \nabla v\|}\right) &= 0, \end{aligned} \quad (3.31)$$

where div is the divergence operator and g' is the first derivative of g. The corresponding Neumann boundary conditions are:

$$\begin{aligned} \frac{g'(\|\nabla u\|)}{\|\nabla u\|}\frac{\partial u}{\partial \mathbf{n}} &= 0 \\ \frac{g'(\|\nabla v\|)}{\|\nabla v\|}\frac{\partial v}{\partial \mathbf{n}} &= 0, \end{aligned} \quad (3.32)$$

where $\mathbf{n}$ is the unit normal vector to the boundary $\partial\Omega$ of the image domain Ω, and $\partial/\partial\mathbf{n}$ is the derivative operator in the direction of $\mathbf{n}$.

For $w \in \{u, v\}$, i.e., where w stands for either of the optical flow components, consider at each point a local orthonormal direct coordinate system (η, ξ) defined by unit vectors $\frac{\nabla w}{\|\nabla w\|}$ and its (counter clockwise) orthogonal unit vector $\left(\frac{\nabla w}{\|\nabla w\|}\right)^{\perp}$. In this reference system, the divergence terms in Eq. (3.31) are written:

$$\mathrm{div}\left(\frac{g'(\|\nabla w\|)}{\|\nabla w\|}\nabla w\right) = \frac{g'(\|\nabla w\|)}{\|\nabla w\|}w_{\xi\xi} + g''(\|\nabla w\|)w_{\eta\eta} \quad (3.33)$$

In a region where w is homogeneous, i.e., where $\|\nabla w\|$ is small, we want g to allow smoothing in both orthogonal directions η and ξ, and in the same manner (isotropy). Considering Eqs. (3.31) and (3.33), the conditions to impose are:

$$\begin{aligned} &\lim_{s\to 0} g''(s) = g''(0) > 0 \\ &\lim_{s\to 0} \tfrac{g'(s)}{s} = g''(0) \end{aligned} \quad (3.34)$$

At the limit when $\|\nabla w\| \to 0$, we have:

$$\mathrm{div}\left(\frac{g'(\|\nabla w\|)}{\|\nabla w\|}\nabla w\right) = g''(0)(w_{\eta\eta} + w_{\xi\xi}) = g''(0)\nabla^2 w$$

Therefore, the Euler-Lagrange equations in this case, when $\|\nabla w\| \rightarrow 0$, would be:

$$
\begin{aligned}
I_x(I_x u + I_y v + I_t) &= \tfrac{\lambda}{2} g''(0)\nabla^2 u \\
I_y(I_x u + I_y v + I_t) &= \tfrac{\lambda}{2} g''(0)\nabla^2 v,
\end{aligned} \tag{3.35}
$$

with Neumann boundary conditions

$$
\frac{\partial u}{\partial \mathbf{n}} = 0, \quad \frac{\partial u}{\partial \mathbf{n}} = 0. \tag{3.36}
$$

These equations are those of the Horn and Schunck formulation, which is what we want.

When ∇w is large, as it would be at motion boundaries, we want to smooth w along ξ but inhibit smoothing in the orthogonal direction, i.e., along η. The conditions to set are:

$$
\begin{aligned}
&\lim_{s \to \infty} g''(s) = 0 \\
&\lim_{s \to \infty} \frac{g'(s)}{s} = \beta > 0,
\end{aligned} \tag{3.37}
$$

and the divergence term at the limit when $\|\nabla w\| \rightarrow \infty$ would be:

$$
\operatorname{div}\left(\frac{g'(\|\nabla w\|)}{\|\nabla w\|}\nabla w\right) = \beta w_{\xi\xi}. \tag{3.38}
$$

However, both conditions in Eq. (3.37) cannot be satisfied simultaneously [37, 38]. Instead, the following weaker conditions can be imposed:

$$
\begin{aligned}
&\lim_{s \to \infty} g''(s) = \lim_{s \to \infty} \frac{g'(s)}{s} = 0 \\
&\lim_{s \to \infty} \frac{g''(s)}{\frac{g'(s)}{s}} = 0.
\end{aligned} \tag{3.39}
$$

Accordingly, diffusion is inhibited in both directions at the limit, when $\|\nabla w\| \rightarrow \infty$, but is otherwise dampened more in direction η than ξ, i.e, smoothing will be dampened more across motion boundaries than along. There are several functions satisfying conditions Eqs. (3.34) and (3.39), $g(s) = 2\sqrt{1+s^2} - 2$ (Aubert), for instance, and the ones shown in Table 3.1.

Table 3.1 Boundary preserving functions for the estimation of optical flow

	g(s)
Aubert	$2\sqrt{1+s^2} - 2$
Geman and Reynolds	$\frac{s}{1+s^2}$
Perona-Malik	$\log(1+s^2)$
Green	$2\log(\cosh(s))$

A discretization of the Euler-Lagrange equations gives a large scale sparse system of nonlinear equations. Instead of solving directly this system, the study in [37, 38] proposed a more efficient implementation using the half-quadratic minimization algorithm applied to the following functional, the change from the original functional being justified by a duality theorem [38]:

$$
\begin{aligned}
E(u, v, b_1, b_2) = & \int_{\Omega} (I_x u + I_y v + I_t)^2 dxdy \\
& + \lambda \int_{\Omega} \left(b_1 \|\nabla u\|^2 + b_2 \|\nabla v\|^2 + \psi(b_1) + \psi(b_2) \right) dxdy \quad (3.40)
\end{aligned}
$$

Two new functions, $b_1(x, y)$ and $b_2(x, y)$, called auxiliary variables, appear in this functional. Also appearing is a function ψ, convex and decreasing, related implicitly to g and such that, for every fixed s, the value of b which minimizes $bs^2 + \psi(b)$ is given by

$$
b = \frac{g'(s)}{2s} \quad (3.41)
$$

This result is the basis of the half-quadratic greedy minimization algorithm which, after initialization, repeats two consecutive steps until convergence. Each iteration performs a minimization with respect to u, v with b_1, b_2 assumed constant followed by a minimization with respect to b_1, b_2 with u, v assumed constant.

Minimization with respect to u, v, with b_1, b_2 considered constant, consists of minimizing the following functional:

$$
\int_{\Omega} (I_x u + I_y v + I_t)^2 + \lambda \left(b_1 \|\nabla u\|^2 + b_2 \|\nabla v\|^2 \right) dxdy \quad (3.42)
$$

The corresponding Euler-Lagrange equations are:

$$
\begin{aligned}
I_x (I_x u + I_y v + I_t) &= \lambda \mathrm{div}(b_1 \nabla u) \\
I_y (I_x u + I_y v + I_t) &= \lambda \mathrm{div}(b_2 \nabla v),
\end{aligned} \quad (3.43)
$$

with Neumann boundary conditions $\partial u / \partial \mathbf{n} = \partial v / \partial \mathbf{n} = 0$. Discretization of the equations yields a large scale sparse linear system of equations which can be solved efficiently with the Gauss-Seidel or the Jacobi method. The divergence terms in Eq. (3.43) can be discretized as in [49].

The minimization with respect to b_1, b_2, with u, v considered constant, consists of minimizing the functional:

$$
\int_{\Omega} \left(b_1 \|\nabla u\|^2 + b_2 \|\nabla v\|^2 + \psi(b_1) + \psi(b_2) \right) dxdy \quad (3.44)
$$

The unique solution is given analytically following Eq. (3.41):

$$
\begin{aligned}
b_1 &= \frac{g'(\|\nabla u\|)}{2\|\nabla u\|} \\
b_2 &= \frac{g'(\|\nabla v\|)}{2\|\nabla v\|}
\end{aligned}
\tag{3.45}
$$

The half-quadratic algorithm to minimize Eq. (3.40) can be summarized as follows:

1. Initialize b_1, b_2
2. Repeat until convergence

   a. Minimize with respect to u, v using Jacobi (or Gauss-Seidel) iterations to solve the linear system of equations corresponding to the discretized Eq. (3.43).
   b. Minimize with respect to b_1, b_2 using Eq. (3.45) $\left[b_1 = \frac{g'(\|\nabla u\|)}{2\|\nabla u\|},\ b_2 = \frac{g'(\|\nabla v\|)}{2\|\nabla v\|}\right]$

Example: This example (courtesy of R.Deriche) uses the Hamburg Taxi sequence of a street intersection scene (from Karlsruhe University, Germany, Institut für Algorithmen und Kognitive Systeme, http://i21www.ira.uka.de/image_sequences/): Fig. 3.4a shows one of the two consecutive images used. The other figures contain a graphical display of the flow field in the rectangular zoom window drawn in (a) (which includes the white car in the center of the intersection and a small portion of the dark-coloured car next to it): Methods of (b) Horn and Schunck, (c) Lucas-Kanade, and (d) Deriche-Aubert-Kornprobst. Visual inspection reveals a motion smoothing spread in (b) and a lack of spatial regularization in (c). In (d) the smooth motion field is well confined to the moving cars as a result of discontinuity preserving smoothness regularization.

Example: This other example uses the synthetic sequence depicted in Fig. 3.5a (*Marbled blocks* sequence from Karlsruhe University, Germany, Institut für Algorithmen und Kognitive Systeme). The camera and the block on the left do not move. The block on the right moves away to the left. The images had noise added. The texture variation is weak at the top edges of the blocks. Depth, and image motion thereof, varies sharply at the blocks boundaries not in contact with the floor. The blocks cast shadows which display apparent motion. The ground truth optical flow vectors are displayed in Fig. 3.5b. Vectors computed with the method of Horn and Schunck and of Deriche, Aubert, Kornprobst are displayed in Fig. 3.5c and d, respectively. The average errors per pixel in magnitude (pixels) and direction (degrees) are (0.142, 5.095) and (0.130, 4.456) for the Horn and Schunck and the Aubert, Deriche, Kornprobst methods, respectively [61]. The better performance of the latter scheme is likely due to the better handling of motion discontinuities as a visual inspection tends to corroborate.

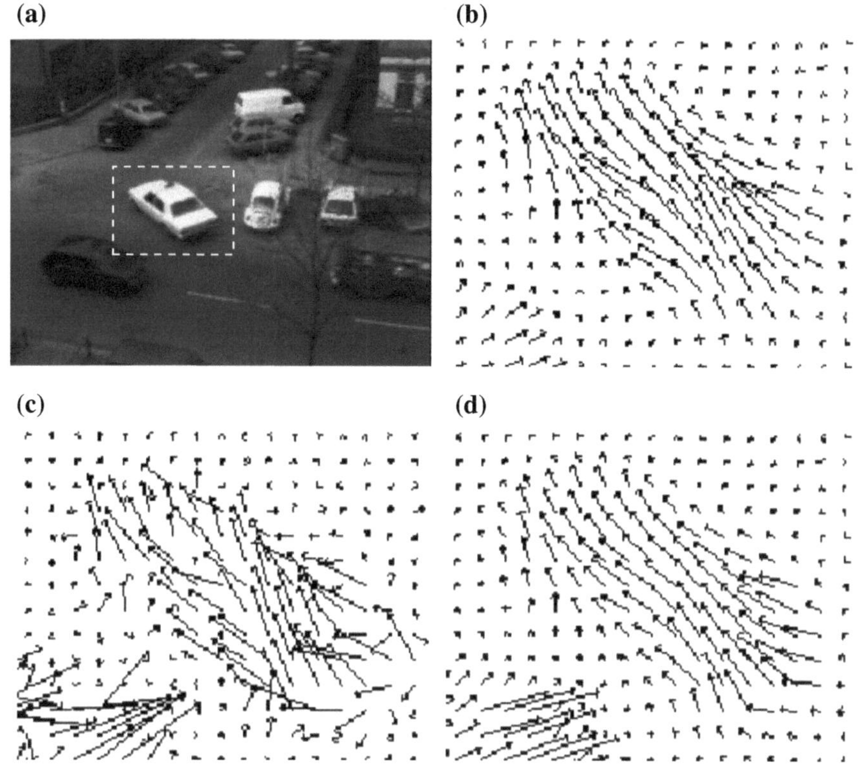

Fig. 3.4 Optical flow estimation on the Hamburg Taxi sequence (courtesy of Deriche, Aubert, and Kornprobst): **a** One of the two consecutive images used. A graphical display of the flow field in the rectangular window shown in **a**, by the methods of **b** Horn and Schunck, **c** Lucas-Kanade, and **d** Deriche, Aubert, and Kornprobst. This last method produces a smooth field confined to the moving objects (the white car and part of the dark car) as a result of discontinuity preserving smoothness regularization

3.6 Image-Guided Regularization

Consider from a probabilistic viewpoint the problem of estimating optical flow from two consecutive images I_1 and I_2. This consists of maximizing the posterior probability $P(W|I_1, I_2)$ over the space of all possible optical flow fields W. This probability is proportional to the product $P(I_2|W, I_1)P(W|I_1)$. The first term, $P(I_2|W, I_1)$, is a term of conformity of W to the data because it is the likelihood that connects I_2 to I_1 via W. The second term, $P(W|I_1)$, is a prior on W which exhibits a partial dependence on data through the conditioning on I_1. This dependence is often ignored and the conditioning on I_1 is removed, resulting in a prior independent of any observation. This is equivalent to imposing statistical independence of W and I_1. However, the dependence is genuine because motion edges often occur at image intensity edges [27]. Therefore, its inclusion in a prior, or a regularization term in

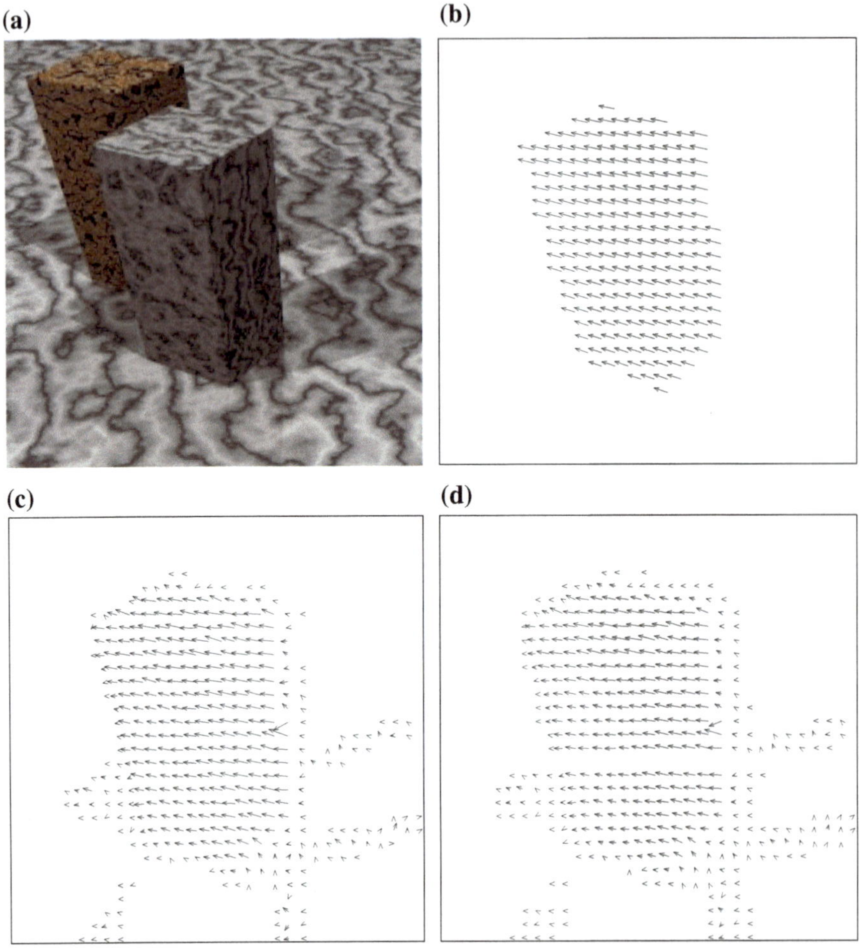

Fig. 3.5 Optical flow estimation on the Marbled block sequence: **a** the first of the two images used, **b** ground truth optical flow, and optical flow by the method of **c** Horn and Schunck, **d** Deriche, Aubert, Kornprobst. This last method produces a smooth field confined to the moving objects as a result of discontinuity preserving smoothness regularization

energy based formulations, affords the opportunity to smooth the motion field without blurring its boundaries by allowing smoothing along the isophote, i.e., in the direction perpendicular to the image spatial gradient, and inhibiting or dampening it across. This can be done via an appropriate gradient-dependent linear transformation $\mathbf{A}(\nabla I)$ of the motion field in the prior/regularization term. Here following are two possible formulations [16, 19, 20].

3.6.1 The Oriented-Smoothness Constraint

The following functional was investigated in [16, 86]:

$$\mathscr{E}(u,v) = \int_{\Omega} (I_1(x-u, y-v) - I_2(x,y))^2 \, dxdy \tag{3.46}$$

$$+ \lambda \int_{\Omega} \left(\nabla u^T \mathbf{A}(\nabla I_1) \nabla u + \nabla v^T \mathbf{A}(\nabla I_1) \nabla v \right) dxdy. \tag{3.47}$$

Matrix $\mathbf{A}$ is defined as a function of the image partial derivatives by:

$$\mathbf{A}(\nabla I_1) = \frac{1}{\|\nabla I_1\|^2 + 2\mu^2} \left[\begin{pmatrix} I_{1y} \\ -I_{1x} \end{pmatrix} (I_{1y} \ \ -I_{1x}) \ + \ \mu^2 \mathbf{I} \right], \tag{3.48}$$

where $\mathbf{I}$ is the identity matrix and μ a constant. The functional was later modified [87] to remove the peculiarity that motion is applied to I_2 in the data term but to I_1 in the regularization term.

An analysis in [20] determined an image-guided regularization matrix $\mathbf{A}$ by imposing on it conditions which would cause smoothing along intensity edges but dampening it across. The analysis is as follows.

3.6.2 Selective Image Diffusion

Consider the following objective functional:

$$\mathscr{E}(W) = \int_{\Omega} (I_x u + I_y v + I_t)^2 dxdy + \lambda \int_{\Omega} (\|\mathbf{A}\nabla u\|^2 + \|\mathbf{A}\nabla v\|^2) dxdy, \tag{3.49}$$

where $\mathbf{A} = \mathbf{A}(\nabla I)$ is a 2×2 matrix which depends on the image structure via the image spatial gradient. Matrix $\mathbf{A}$ must be chosen so as to allow smoothing at each point in the direction of the perpendicular to the image gradient, i.e., along the isophote, and dampen it in the direction of the gradient, i.e., perpendicular to the isophote. This can be done by imposing the following conditions on the eigenvalues α_1, α_2 of $\mathbf{A}$ [20]:

1. For $\|\nabla I\| \neq 0$, $\mathbf{x}_1 = \frac{\nabla I}{\|\nabla I\|}$, $\mathbf{x}_2 = \left(\frac{\nabla I}{\|\nabla I\|}\right)^{\perp}$ are the unit eigenvectors corresponding to α_1, α_2,
2. $\alpha_2 = 1$
3. α_1 is a monotonically decreasing continuous function of $\|\nabla I\|$ such that: $\lim_{\|\nabla I\| \to 0} \alpha_1 = 1$ and $\lim_{\|\nabla I\| \to \infty} \alpha_1 = 0$.

Intuitively, the purpose of these conditions is as follows: the first condition says that the two orthogonal directions which should be considered for smoothing are those of the isophote and the image gradient; the second condition is to allow full smoothing along the isophote; the third condition stipulates that smoothing along the gradient direction is to be allowed only to a degree that decreases with the intensity edge strength, varying from full to no strength.

The Euler-Lagrange equations corresponding to Eq. (3.49) are:

$$\begin{aligned} I_x(I_x u + I_y v + I_t) - \lambda \mathrm{div}(\mathbf{B}\nabla u) &= 0 \\ I_y(I_x u + I_y v + I_t) - \lambda \mathrm{div}(\mathbf{B}\nabla v) &= 0, \end{aligned} \tag{3.50}$$

where $\mathbf{B} = \mathbf{A}^t\mathbf{A} + \mathbf{A}\mathbf{A}^t$, with Neumann boundary conditions:

$$\begin{aligned} \mathbf{B}\nabla u \cdot \mathbf{n} &= 0 \\ \mathbf{B}\nabla v \cdot \mathbf{n} &= 0 \end{aligned} \tag{3.51}$$

Let $\mathbf{P}$ be the 2×2 orthogonal matrix $\mathbf{P} = (\mathbf{x}_1, \mathbf{x}_2)$, i.e., whose columns are $\mathbf{x}_1$ and $\mathbf{x}_2$, and let

$$\Lambda = \begin{pmatrix} \alpha_1 & 0 \\ 0 & \alpha_2 \end{pmatrix} \tag{3.52}$$

Using the first condition, we have, by definition, $\mathbf{AP} = \mathbf{P}\Lambda$. Therefore, $\mathbf{A} = \mathbf{P}\Lambda\mathbf{P}^{-1} = \mathbf{P}\Lambda\mathbf{P}^t$, since $\mathbf{P}$ is orthogonal. This gives, using the second condition $(\alpha_2 = 1)$,

$$\mathbf{A}(\nabla I) = \frac{1}{\|\nabla I\|^2} \begin{pmatrix} \alpha_1 I_x^2 + I_y^2 & (\alpha_1 - 1) I_x I_y \\ (\alpha_1 - 1) I_x I_y & I_x^2 + \alpha_1 I_y^2 \end{pmatrix} \tag{3.53}$$

Using the following α_1, which satisfies the third condition,

$$\alpha_1 = \frac{1}{(1 + \frac{\|\nabla I\|^2}{\mu^2})^{\frac{1}{2}}}, \tag{3.54}$$

where μ is a parameter to modulate the strength of smoothing, we have:

$$\mathbf{B} = \frac{1}{\mu^2 \|\nabla I\|^2} \begin{pmatrix} \mu^2 + I_y^2 & -I_x I_y \\ -I_x I_y & \mu^2 + I_x^2 \end{pmatrix} \tag{3.55}$$

Assuming $\|\nabla I\|$ is bounded on Ω, this matrix is positive definite, which means that Eq. (3.50) are diffusion equations. To see intuitively that they realize the desired diffusion, note that where $\|\nabla I\|$ is small, α_1 is close to 1 and, therefore, $\mathbf{A}(\nabla I)$ is close to the identity, causing the regularization term in Eq. (3.49) to be close to the L^2 norm and Eq. (3.50) to behave isotropically. When, instead, $\|\nabla I\|$ is large, $\mathbf{A}(\nabla I)$ approximates a projection onto the direction perpendicular to ∇I and only the projection of ∇u and ∇v along that direction will contribute to the regularization.

Another way to see that we have the desired diffusion is by looking at the behaviour of Eq. (3.50) locally at each point, in a small neighborhood where ∇I is constant and nonzero. In this neighborhood, consider the local coordinate system (η, ξ) according to the reference system defined by $\frac{\nabla I}{\|\nabla I\|}, \left(\frac{\nabla I}{\|\nabla I\|}\right)^{\perp}$. In this orthonormal reference system, we have $\nabla u = (u_\eta, u_\xi)$ and $\nabla v = (v_\eta, v_\xi)$, and

$$\mathbf{B} = \begin{pmatrix} \alpha_1^2 & 0 \\ 0 & \alpha_2^2 \end{pmatrix}. \tag{3.56}$$

which gives the following local form of the divergence term, using $\alpha_2 = 1$:

$$\mathrm{div}(\mathbf{B}\nabla u) = \alpha_1^2 u_{\eta\eta} + u_{\xi\xi} \tag{3.57}$$

and, therefore, the local form of the Euler-Lagrange equations:

$$\begin{aligned} I_x(I_x u + I_y v + I_t) - \lambda(\alpha_1^2 u_{\eta\eta} + u_{\xi\xi}) = 0 \\ I_y(I_x u + I_y v + I_t) - \lambda(\alpha_1^2 v_{\eta\eta} + v_{\xi\xi}) = 0 \end{aligned} \tag{3.58}$$

It is clear from these equations that diffusion will occur along axis ξ, i.e, along the intensity edge and that it will be dampened along axis η, i.e., along the direction of the gradient. Since α_1 is a decreasing function of $\|\nabla I\|$, the degree of dampening will be commensurate with the edge strength. Parameter μ in Eq. (3.54), although not essential, can be used to control how fast with respect to edge strength dampening occurs across edges.

The minimization of Eq. (3.49) can be done by the corresponding Euler-Lagrange descent equations [20], namely,

$$\begin{aligned} \frac{\partial u}{\partial \tau} &= -I_x(I_x u + I_y v + I_t) + \lambda \mathrm{div}(\mathbf{B}\nabla u) \\ \frac{\partial v}{\partial \tau} &= -I_y(I_x u + I_y v + I_t) + \lambda \mathrm{div}(\mathbf{B}\nabla v) \end{aligned} \tag{3.59}$$

One can also discretize the Euler-Lagrange equations Eq. (3.50). This would give a large scale sparse system of linear equations which can be solved efficiently by Gauss-Seidel or Jacobi iterations.

3.7 Minimum Description Length

A way to preserve motion discontinuities is to bring in the length of the discontinuity set in the regularization [30]. A boundary length term commonly appears in image segmentation functionals, first in the Mumford and Shah functional [32]. It is essential in the Leclerc's minimum description length (MDL) formulation [33] which we focus on in this section and transpose to optical flow estimation. The Leclerc's method can be viewed as a discrete implementation of the Mumford-Shah functional [88]. It minimizes an objective function which assigns a code length to an image partition described according to a predefined "description language."

Let I_0 be an observed image with discrete domain D and I an approximation corresponding to a partition $\mathscr{R} = \{R_k\}$ of the image domain D into regions where the image is modelled by a parametric model with parameters $\{\theta_k\}$. The Leclerc MDL criterion [33] to estimate the image underlying I_0 is:

$$E(\mathscr{R}, \{\theta_k\}) = \frac{a}{2} \sum_k l_k - \sum_k \sum_{i \in R_k} log_2 P(I_i|\theta_k) + \sum_k b_k, \tag{3.60}$$

where l_k is the length of the boundary of R_k in terms of the number of pixels it threads through, a is the bit cost of coding one edge element, and b_k is the bit cost of coding the (discrete) parameter vector of region R_k. The first term is the code length for the boundaries and the second for the image given the region parameters. The last term is the code length to describe the region models via their parameters; assuming equal code length for all regions, the term can be dropped from the criterion. For a piecewise constant description of I and quantized Gaussian noise, the criterion can be re-written as [33]:

$$E(I) = \frac{a}{2} \sum_{i \in D} \sum_{j \in \mathscr{N}_i} (1 - \delta(I_i - I_j)) + b \sum_{i \in D} \frac{(I_i - I_{0i})^2}{\sigma^2}, \tag{3.61}$$

where $a \approx 2$ and $b = \frac{1}{2log2}$; $\mathscr{N}_i$ is some fixed neighborhood of pixel i; and

$$\delta(z) = \begin{cases} 1 & \text{for } z = 0 \\ 0 & \text{for } z \neq 0 \end{cases} \tag{3.62}$$

Energy Eq. (3.61) can be solved by a *continuation* scheme indexed by the standard deviation of a Gaussian substituted for δ: Starting from an initial large value, the standard deviation is gradually lowered and, at each step, a solution to the corresponding problem is computed using as initial approximation the solution to the previous problem.

A continuum version of the Leclerc's MDL criterion is [89], assuming the code length to describe the parametric models is common to all regions:

$$\mathscr{E}(\Gamma, \{\theta_k\}) = \sum_k \left(\frac{a}{2} \int_{\partial R_k} ds - log P(\{I(\mathbf{x}) : \mathbf{x} \in R_k\}|\theta_k) \right), \tag{3.63}$$

where $\{R_k\}$ is a partition of the image domain Ω, $\Gamma = \{\partial R_k\}$ its boundaries, and $\{\theta_k\}$ the regions parameters. The code length to describe the parametric models was assumed common to all regions and has been been dropped. A transposition to optical flow can be written:

$$\mathscr{E}(\Gamma, \{\theta_k\}) = \sum_k \left(\frac{a}{2} \int_{\partial R_k} ds - log P(\{r(\mathbf{x})\} : \mathbf{x} \in R_k | \theta_k) \right), \tag{3.64}$$

where $r(\mathbf{x}) = (I_x u + I_y v + I_t)(\mathbf{x})$. If we assume independent identical probability models for r everywhere on Ω, then Eq. (3.64) can be re-written:

$$\mathscr{E}(\Gamma, \{\theta_k\}) = \frac{a}{2} \sum_k \int_{\partial R_k} ds - \int_{\Omega} log P(r(\mathbf{x})) d\mathbf{x}). \tag{3.65}$$

A discretization of the length term is:

$$\frac{a}{2} \sum_{i \in D} \sum_{j \in \mathcal{N}_i} \left(1 - \delta(u_i - u_j)\delta(v_i - v_j)\right), \tag{3.66}$$

where $a \approx 2$. For r normally distributed with variance σ^2, a discretization of the data term can be written:

$$c + b \sum_{i \in D} \frac{(I_{xi} u_i + I_{yi} v_i + I_{ti})^2}{\sigma^2}, \tag{3.67}$$

where $b = \frac{1}{2 \log 2}$ and c is a constant to ignore [33]. The minimum description length estimate of optical flow is the motion field $\tilde{W}$ over D which corresponds to a minimum of the total code length of description:

$$E(W) = b \sum_{i \in D} \frac{(I_{xi} u_i + I_{yi} v_i + I_{ti})^2}{\sigma^2} + \frac{a}{2} \sum_{i \in D} \sum_{j \in \mathcal{N}_i} \left(1 - \delta(u_i - u_j)\delta(v_i - v_j)\right). \tag{3.68}$$

Numerical Implementation

The objective function Eq. (3.68) is not differentiable due to the presence of the delta function, as in the Leclerc objective function for intensity images. This suggests to embed the minimization of Eq. (3.68) in a family of minimizations indexed by the parameter of a differentiable approximation of the δ function, and use *continuation* [33, 81] to carry out the estimation. Continuation can be based on the following substitution:

$$\delta(u_i - u_j)\delta(v_i - v_j) \quad \leftarrow \quad e_{ij}(W, s) = e^{-\frac{(u_i - u_j)^2 + (v_i - v_j)^2}{(s\sigma)^2}} \tag{3.69}$$

Using $s\sigma$ in Eq. (3.69), rather that s, simplifies subsequent expressions without causing a loss of generality. The actual parameter of continuation remains s. With substitution Eq. (3.69), the objective function to minimize is re-written:

$$E(W, s) = b \sum_{i \in D} \frac{(I_{xi}u_i + I_{yi}v_i + I_{ti})^2}{\sigma^2} + \frac{a}{2} \sum_{i \in D} \sum_{j \in \mathcal{N}_i} (1 - e_{ij}(W, s)) \tag{3.70}$$

Let $s_1, s_2, ...$ be a decreasing sequence of s values tending to zero. Continuation solves the following sequence of problems indexed by these values of s:

$$\text{Minimize} \quad E(W, s_l) \tag{3.71}$$

For each value s_l of s, the necessary conditions for a minimum of E give two constraints at each $i \in D$:

$$\begin{aligned} I_{xi}(I_{xi}u_i + I_{yi}v_i + I_{ti}) + a_l \textstyle\sum_{j \in \mathcal{N}_i} (u_i - u_j) e_{ij}(W, s_l) = 0 \\ I_{yi}(I_{xi}u_i + I_{yi}v_i + I_{ti}) + a_l \textstyle\sum_{j \in \mathcal{N}_i} (v_i - v_j) e_{ij}(W, s_l) = 0, \end{aligned} \tag{3.72}$$

where $a_l = a \log 2/s_l^2$. This yields a large scale sparse system of equations most of which are linear, and that can be solved using the following Jacobi-type iterative scheme where each iteration applies a Jacobi update to a linear system of equations obtained by evaluating the exponential term with the values of motion computed at the preceding iteration:

$$\begin{aligned} u_i^{k+1} &= \frac{-I_{x_i} I_{t_i} - I_{x_i} I_{y_i} v_i^k + a_l \sum_{j \in \mathcal{N}_i} e_{ij}^k(W, s_l) u_j^k}{I_{x_i}^2 + a_l \sum_{j \in \mathcal{N}_i} e_{ij}^k(W, s_l)} \\ v_i^{k+1} &= \frac{-I_{y_i} I_{t_i} - I_{x_i} I_{y_i} u_i^k + a_l \sum_{j \in \mathcal{N}_i} e_{ij}^k(W, s_l) v_j^k}{I_{y_i}^2 + a_l \sum_{j \in \mathcal{N}_i} e_{ij}^k(W, s_l)} \end{aligned} \tag{3.73}$$

The solution of each problem in Eq. (3.71) serves as the initial solution for the next problem. As s approaches zero, the problem approaches the original Eq. (3.68) because $e_{ij}(s)$ tends to δ. When s tends to ∞, the second term in Eq. (3.70) approaches 0. This suggest that the first problem be stated with s large, using, for instance, the normal component vector of optical flow as initial approximation. The iterations are continued up to a small s_l. As a rule of thumb, about 100 iterations of continuation and 5 of Eq. (3.73) sufficed in experiments.

Example: The MDL estimation scheme is illustrated using the Marbled blocks synthetic test sequence (*Marmor-2* sequence from the KOGS/ IAKS laboratory database, University of Karlsruhe, Germany). The rightmost block moves away to the left and the small center block forward to the left. The camera and the leftmost block are static. The images have been noised. The texture variation is weak at the top edges of the blocks. Depth varies sharply at the blocks boundaries not in contact with the floor. The blocks cast some shadows. The scene and the actual motion field are shown in Fig. 3.6a, and the MDL motion estimate in Fig. 3.6b. In spite of its embedded motion

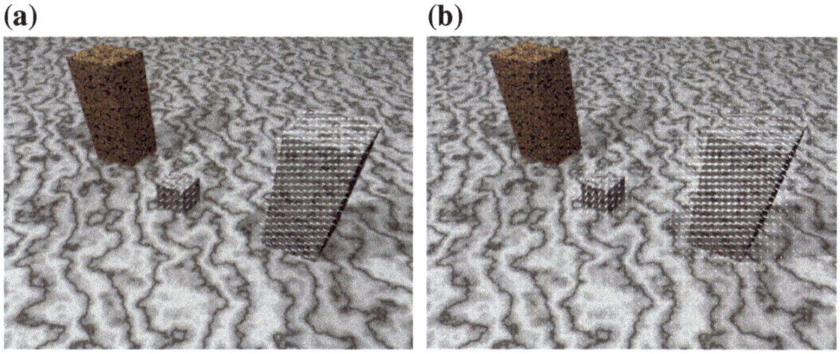

Fig. 3.6 **a** Ground truth image motion superimposed on the first of the two Marbled blocks images used in the experiment and, **b** the MDL motion estimate. In spite of its embedded motion boundary preservation, the scheme has let some smoothing, although mild, across the blocks occluding contours where there are motion discontinuities

boundary preservation, the scheme has let some smoothing, although mild, across the blocks occluding contours where depth, and motion thereof, vary sharply and significantly. The average error on the motion magnitude, over the whole image, is 0.13 pixel and the average direction error on the two moving blocks is 4.7°. The standard deviations are 0.2 for the magnitude, 5.8 for the direction for the small block and 3.5 for the large block. These statistics are comparable to those of the Horn and Shunck and the Deriche, Aubert, and Kornprobst schemes.

3.8 Parametric Estimation

Parametric motion estimation in a support region $R \subset \Omega$ consists of representing the motion field in R by a parametric model and using the spatiotemporal data to determine the parameters which provide the best fit. One of the main motivations for parametric motion estimation is economy of description because motion in the support region R can be compactly described by the set of model parameters. Parametric estimation also forgoes the need for regularization in R because it implies smoothness of motion. We will focus on linear parametric models. They are analytically convenient to use and, when chosen properly, can be powerful so as to represent fine details of motion.

Parametric estimation of optical flow over a support region R can be set up as follows [60]. Let $\theta_j : (x, y) \in \Omega \rightarrow \theta_j(x, y) \in \mathbb{R}, \quad j = 1, ..., M$ be basis functions and L their span: $L = \text{span}\{\theta_1, ..., \theta_M\}$. Each of the coordinate functions u, v of optical flow W is considered an element of L :

$$W = \alpha^T \theta \tag{3.74}$$

where θ is the vector of basis functions:

$$\theta = \left(\theta_1 \; \theta_2 \; \cdots \; \theta_M\right)^T \tag{3.75}$$

and α is the matrix of *parameters*, i.e., of the coefficients of the expansion of motion in the basis of L:

$$\alpha = \begin{pmatrix} \alpha_{11} \; \alpha_{21} \; \cdots \; \alpha_{M1} \\ \alpha_{12} \; \alpha_{22} \; \cdots \; \alpha_{M2} \end{pmatrix}^T \tag{3.76}$$

The first row of α^T has the parameters of u and the second row those of v. The parameters in R are computed by minimizing the following functional which uses the optical flow constraint in which Eq. (3.74) is substituted:

$$\mathscr{E}(\alpha) = \int_R (\nabla I \cdot \alpha^T \theta + I_t)^2 dxdy. \tag{3.77}$$

The corresponding least squares equations to determine the parameters are:

$$\mathbf{B}\beta + \mathbf{d} = \mathbf{0} \tag{3.78}$$

where:

- Vector β is the $2M \times 1$ vector constructed by vertical concatenation of the parameters α_1 and α_2 corresponding to optical flow components u and v:

$$\begin{aligned} \beta[m] &= \alpha_{m1} \\ \beta[M+m] &= \alpha_{m2}, \end{aligned} \tag{3.79}$$

  for $m = 1, \ldots, M$.
- Matrix $\mathbf{B}$ is the following $2M \times 2M$ matrix formed by the vertical and horizontal concatenation of 4 $M \times M$ sub-matrices $\mathbf{B}_{rc}$:

$$\mathbf{B} = \begin{bmatrix} \mathbf{B}_{11} \; \mathbf{B}_{12} \\ \mathbf{B}_{21} \; \mathbf{B}_{22} \end{bmatrix}, \tag{3.80}$$

  where the elements of the sub-matrices are defined by:

$$B_{rc}[m, n] = \int_R I_r I_c \theta_m \theta_n \; dxdy, \tag{3.81}$$

  for $m = 1, \ldots, M$, $n = 1, \ldots, M$ and I_l being the spatial derivative of I in the horizontal ($l = 1$) and vertical ($l = 2$) directions.
- Vector $\mathbf{d}$ is the $2M \times 1$ vector with the following elements, for $m = 1, \ldots, M$:

$$\begin{aligned} d[m] &= \textstyle\int_R I_t I_1 \theta_m \; dxdy \\ d[M+m] &= \textstyle\int_R I_t I_2 \theta_m \; dxdy \end{aligned} \tag{3.82}$$

Region R is the support for the formulas above and the question arises as to which region to use to compute the motion field in Ω. Several possibilities can be envisaged. One can use $R = \Omega$. In this case, the problem would be to choose the basis functions and their number. Large images in which several complex motions occur, independent human motions for instance, are likely to require a large number of parameters, which might invalidate the argument of representation economy and also reduce the effectiveness of the scheme. Another possibility is to formulate the problem as joint parametric motion estimation and segmentation. Segmentation would be a partition $\mathscr{R} = \{R_i\}_1^N$ of Ω into N regions differing by their motion as described by a parametric model, i.e., regions each with its own set of parameters. This problem will be investigated in (Sect. 3.11).

Another way to do parametric motion estimation, which does not resort to least squares fit over Ω or use joint estimation and segmentation, has been investigated in [44]. The scheme, called over-parametrization, uses a set of parameters at each point $(x, y) \in \Omega$, i.e., $\alpha = \alpha(x, y)$ and, showing the dependence of the parameters on position:

$$W(x, y) = \alpha(x, y)^T \theta(x, y) \tag{3.83}$$

A linearized optical flow constraint version of the data term in [44] can be written:

$$\int_\Omega g\left(\nabla I(x, y) \cdot \alpha^T(x, y)\theta(x, y) + I_t(x, y)\right) dxdy, \tag{3.84}$$

where $g(z) = \sqrt{z^2 + \varepsilon^2}$, for some small ε, which induces an approximate L^1 metric. The idea of over-parametrization was also used in image segmentation by Leclerc's MDL scheme [33] which looked at an image as a position-dependent parametric function of position. The constant and polynomial models were explicitly treated. Leclerc used the length of the motion boundary set to regularize the parameter field. This set is evaluated in the MDL cost by explicitly defining an edge to be a point between two regions of differing parametric motion descriptions. In [44], the regularization acts directly on the parameters and has the form:

$$\int_\Omega g\left(\sum_{i=1}^{2}\sum_{j=1}^{M} \|\alpha_{ij}\|^2\right). \tag{3.85}$$

Alternatively, it may be appropriate to use a boundary-preserving function of the type we discussed earlier. As with the boundary length term of Leclerc MDL formulation, this regularization implies that regions formed by functional minimization are characterized by one set of motion parameters and regions differ from each other by this set.

Let δ be the optical flow parametric representation data term:

$$\delta = \nabla I \cdot \alpha^T \theta + I_t. \tag{3.86}$$

The Euler-Lagrange equations corresponding to the minimization of the over-parametrization functional:

$$\int_{\Omega} g\left(\delta^2\right) dxdy + \lambda \int_{\Omega} g\left(\sum_{i=1}^{2}\sum_{j=1}^{M} \|\alpha_{ij}\|^2\right) dxdy \tag{3.87}$$

are given by, for $j = 1, \ldots, M$:

$$\begin{aligned} g'\left(\delta^2\right)\delta I_x \alpha_{1j} + \lambda div\left(g'\left(\sum_{i=1}^{2}\sum_{j=1}^{M}\|\alpha_{ij}\|^2\right)\nabla\alpha_{1j}\right) &= 0 \\ g'\left(\delta^2\right)\delta I_y \alpha_{2j} + \lambda div\left(g'\left(\sum_{i=1}^{2}\sum_{j=1}^{M}\|\alpha_{ij}\|^2\right)\nabla\alpha_{2j}\right) &= 0 \end{aligned} \tag{3.88}$$

The formulation can be generalized to use the displaced frame difference in the data term rather than its Horn and Schunck linearized form [44, 90]. The equations are nonlinear. An efficient numerical implementation, within multiresolution processing (Sect. 3.10), is described in [44], with a validation experiment using the Yosemite test image sequences.

3.9 Variations on the Data and Smoothness Terms

To preserve motion boundaries some studies have used the L^1-norm for the optical flow smoothness term of the objective functional [91, 44, 92], in lieu of the quadratic regularization term of Horn and Schunck [10]. However, there has been no analysis or experimentation to support a comparison of the L^1 norm and discontinuity preserving functions of the type in Table 3.1, the Aubert function for instance. The L^1 norm has also been considered for the data term, to evaluate the displaced frame difference, or its continuous total temporal derivative expression. However, there is some evidence from an investigation of temporal noise in image sequences [93] that the L^2 norm may be more appropriate.

Data functions other than the displaced frame difference, or its total temporal derivative continuous expression, have been suggested and some have been investigated experimentally [42], for instance those which arise from the invariance along motion trajectories of the image gradient or of its norm, the norm of the Laplacian, and the norm or trace of the Hessian. Some of these variations have exhibited very accurate results on the Yosemite test sequences.

3.10 Multiresolution and Multigrid Processing

Multiresolution and multigrid processing are "multilevel" computations which solve a system of equations on a given discretization grid by solving similar systems on grids at coarser discretizations. Although conceptually similar from this general point of view, multiresolution and multigrid processing differ in the order they visit the coarser grids and in the type of variables they compute at each of these grids.

3.10.1 Multiresolution Processing

The optical flow constraint, which enters the formulations we have discussed, uses the image temporal derivative, i.e., the image rate of change along the temporal axis. In practice, of course, we have to estimate the motion field from a discrete-time image sequence and if velocities are large, typically to cause displacements of over a pixel between consecutive views, the image temporal derivative may not be approximated sufficiently accurately to bear on velocity estimation, even when the image spatial resolution is high. In such a case, motion estimation can benefit from *multiresolution* processing [51, 22, 94, 28, 95]. In this coarse-to-fine strategy, estimation is served by a pyramidal image representation [96] in which an image is processed by filtering-and-subsampling into a pyramid of images of successively lower resolution. The original image is at the base of the pyramid. As motion extent decreases with resolution, the goal is to start processing at a pyramid level where this extent is within range of estimation. The estimate at this level is then projected on the level immediately below to warp the image at this level. The warped image is processed for an increment of motion, also assumed within range of estimation, and the scheme is repeated down to the original image at the base of the pyramid. Several variants of this basic coarse-to-fine scheme have been investigated but these have the same driving concepts, as just described, and differ mainly in the way the various steps are accomplished. Black's thesis [22] contains an introductory review of the subject. An actual use of multiresolution processing within a thorough motion estimation framework is given in [51, 28, 95].

The following algorithmic steps show the basic concepts involved in multiresolution estimation of optical flow. First, a pyramid of images is constructed from each of the two original images I_1 and I_2 used in the estimation, by repeated low-pass filtering, with a Gaussian, for instance, and subsampling at a rate of 2:

$$I_j^{l-1}\left(\frac{\mathbf{x}}{2}\right) = h * I_j^l(\mathbf{x}) \quad j = 1, 2, \tag{3.89}$$

where $\mathbf{x} = (x, y)$, l designates the resolution level, corresponding to image size $2^l \times 2^l$ (we assume that the length and width of the original image are powers of 2), the coarsest level being $l = 0$, h is the filter and $*$ designates convolution. Subsampling brings down the optical flow magnitude by a factor of two, the purpose being to have a valid discrete representation of the optical velocity components u and v. The intended purpose of low pass filtering is to bring the wavelength of the image spatial frequency components below the motion magnitude so as to have a valid discrete evaluation of the image temporal derivative. Both operations, filtering and subsampling, concur to make continuous formulations applicable at a pyramid level high enough, i.e., at an image resolution low enough. Optical flow is estimated at this coarsest level and estimation is continued down successive pyramid levels, i.e., up successively higher image resolution, using three basic operations at each level: (1) *prolongation* of the optical flow from the immediately preceding (higher, coarser) level, (2) transformation, at this level, of the first of the two images by this projected flow, called image *warping*, and (3) estimation, at this level, of an *incremental flow* using the warped image and the second image:

At coarsest level $l = 0$ initialize flow field W^0
From $l = 1$ to L

1. Prolong the flow field: $W^l = p(W^{l-1})$
2. Displace (warp) the first image by the flow field: $I_1^l(\mathbf{x}) \leftarrow I_1^l(\mathbf{x} + W)$
3. Estimate the flow increment δW^l from I_1^l and I_2^l
4. Update the flow field: $W^l \leftarrow W^l + \delta W^l$

Prolongation is a coarse-to-fine interpolation operator which assigns to each fine-level point a value interpolated from the values at neighboring coarse-level points, i.e., it fills in the empty grid positions at level l by interpolating neighboring flow values at level $l - 1$ multiplied by 2. The prolongation is generally called *projection* in the optical flow literature although the appellation is discordant with the common mathematical usage of the term. As well, image displacement (warping) at any level is done by interpolating the non-grid (displaced) values of the image.

3.10.2 Multigrid Computation

Multigrid procedures have been used in optical flow estimation [52] generally to complement mutiresolution processing at each level of the image pyramid [97, 95, 67, 98]. Multigrid schemes have had a great impact in numerical analysis where they were developed, particularly to solve iteratively large scale linear systems of equations in boundary value problems for partial differential equations [99, 100]. The main reason for using the multigrid computation is to refine via coarser grids a fine grid approximate solution cheaper, faster, and more accurately than using only the fine grid.

The multigrid method is better explained with (large) systems of linear equations although it is also applicable to nonlinear equations. Let $\mathbf{A}^h\mathbf{z}^h = \mathbf{b}^h$ be a fine-grid system of linear equations, h designating the domain grid spacing. Let $\tilde{\mathbf{z}}^h$ be an approximate solution and $\mathbf{r}^h = \mathbf{b}^h - \mathbf{A}^h\tilde{\mathbf{z}}^h$, called the *residual*. The *error* $\mathbf{e}^h = \mathbf{z}^h - \tilde{\mathbf{z}}^h$ then satisfies:

$$\mathbf{A}^h\mathbf{e}^h = \mathbf{r}^h \tag{3.90}$$

This equation can be transferred to a coarser grid with spacing H, double the spacing for instance, $H = 2h$, as:

$$\mathbf{A}^H\mathbf{e}^H = \mathbf{R}_h^H\mathbf{r}^h, \tag{3.91}$$

where $\mathbf{A}^H$ is a coarse-grid approximation of the fine-grid $\mathbf{A}^h$ and $\mathbf{R}_h^H$ is a *restriction* operator from the fine to the coarse grid, which assigns to each point of the coarse grid some weighed average of its argument evaluated at the neighboring fine-grid points. An approximate solution $\tilde{\mathbf{e}}^H$ of Eq. (3.91) is then computed to *correct* the fine-grid approximation $\tilde{\mathbf{z}}^h$:

$$\tilde{\mathbf{z}}^h \leftarrow \tilde{\mathbf{z}}^h + \mathbf{P}_H^h \tilde{\mathbf{e}}^H, \tag{3.92}$$

where $\mathbf{P}_H^h$ is a coarse-to-fine interpolation operator, also called prolongation, which assigns to each fine-grid point a value interpolated from the values at neighboring coarse-grid points. This basic two-level process is summarized by the following steps [100]:

1. Compute approximate solution $\tilde{\mathbf{z}}^h$ by iterating a few times on $\mathbf{A}^h \mathbf{z}^h = \mathbf{b}^h$
2. Compute fine grid residual $\mathbf{r}^h = \mathbf{b}^h - \mathbf{A}^h \tilde{\mathbf{z}}^h$
3. Restrict $\mathbf{r}^h$ to coarse grid residual $\mathbf{r}^H$ by $\mathbf{r}^H = \mathbf{R}_h^H \mathbf{r}^h$
4. Solve $\mathbf{A}^H \mathbf{e}^H = \mathbf{r}^H$ for error $\tilde{\mathbf{e}}^H$
5. Prolong $\tilde{\mathbf{e}}^H$ to fine grid error $\tilde{\mathbf{e}}^h$ by $\tilde{\mathbf{e}}^h = \mathbf{P}_H^h \tilde{\mathbf{e}}^H$
6. Correct $\tilde{\mathbf{z}}^h$ by $\tilde{\mathbf{z}}^h \leftarrow \tilde{\mathbf{z}}^h + \tilde{\mathbf{e}}^h$
7. Iterate a few times on $\mathbf{A}^h \mathbf{z}^h = \mathbf{b}^h$ from $\tilde{\mathbf{z}}^h$

For common problems, there are standard restriction and prolongation operators, and the coarse-grid version $\mathbf{A}^H$ can be computed from these as $\mathbf{A}^H = \mathbf{R}_h^H \mathbf{A}^h \mathbf{P}_H^h$ [100]. The two-level algorithm above can be extended to a pyramid of more levels by using a hierarchy of coarse grids, for instance with spacings $h, 2h, 4h, \ldots, H$, and calling the two-level algorithm *recursively* at each level except the coarsest, i.e., step 4 of the two-level algorithm is now a recursive call to it except at the coarsest grid where the error is computed to trigger an upward string of error prolongations and corresponding solution corrections. This "deep" V-path is illustrated in Fig. 3.7b.

The multigrid method is essentially different from the multiresolution in that it is an *error estimation* scheme which successively refines an initial approximate solution on the original high-resolution discretization grid using errors calculated on successively coarser grids. Multiresolution computations, instead, refine an initial approximation solved on the coarsest grid by working successively through higher resolutions up to the original fine grid. From this perspective, a multiresolution

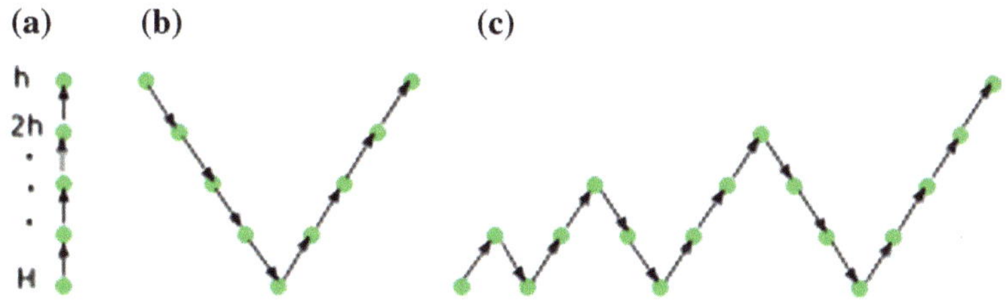

Fig. 3.7 Multiresolution and multigrid paths: **a** Multiresolution processing proceeds from low resolution to high; **b** V-cycle multigrid computations start at the original finest resolution grid to move though successively coarser grids to the coarsest and then up though successively finer grids until the finest; **c** Full multigrid cycle links several V-cycles of different size and the same depth starting at the coarsest grid

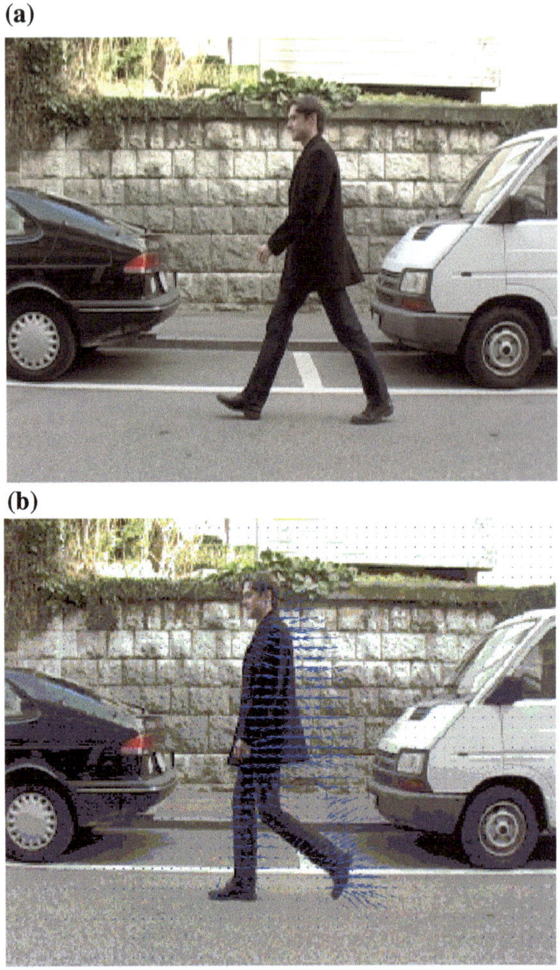

Fig. 3.8 **a** The first of the two images used to compute optical flow; **b** the second image and the flow estimated by embedding in multiresolution processing. The flow occurs, predictably, in both the region uncovered by motion in the first image and the region covered in the second image. Multiresolution computations have been able to capture well the overall movement of the person

scheme adopts a coarse-to-fine strategy, i.e., after creating the image pyramid by low-pass filtering and sub-sampling, it works strictly down the image pyramid, i.e., from lowest resolution to highest (Fig. 3.7a), whereas multigrid processing moves both ways in this pyramid, first down to successively coarser resolutions and then back up to successively finer resolutions up to the original to apply corrections computed from an error solved at the coarsest resolution. Several of these V-shaped paths of different sizes but of the same depth can be linked into a string that starts at the coarsest grid to give the *full multigrid cycle*. This is illustrated in Fig. 3.7c. Nonlinear equations are generally resolved with such a cycle of computations.

Example: It is remarkable that multiresolution/multigrid processing works at all when displacements between views are significantly large as in the following example. The displacements in the two images used here, of a person walking, are quite large. The first image is shown in Fig. 3.8a and the second in Fig. 3.8b which also displays the optical flow estimated by embedding in multiresolution processing. What should be pointed out in this example is that the flow seems to capture well the overall movement of the person in spite of the large displacements. Predictably, motion is found and estimated in both the region unveiled by motion in the first image and the region covered by motion in the second image. This kind of result can serve motion detection as we will see in Chap. 4.

3.11 Joint Estimation and Segmentation

Segmentation, or partitioning, of the flow field with concurrent estimation within each segmentation region with no particular concern about motion boundaries is an alternative to estimation with boundary-preserving regularization because segmentation will place boundaries between regions of significantly differing motion, therefore at significant flow edges. The usefulness of joint optical flow estimation and segmentation by variational methods was first investigated in [25, 101]. Active contours as motion boundary variables were used [54, 60, 102, 103], and embedding three-dimensional rigid body interpretation in estimation was considered in [58, 61]. When motion-based image partitioning is the main purpose of concurrent flow field estimation and segmentation, a coarse model of image motion such as piecewise constant or affine can be sufficient, particularly when this motion is due to viewing system movement and rigid environmental objects. However, given a flow-based segmentation obtained with a coarse motion model, one can apply a more accurate optical flow algorithm a posteriori in each segmentation region separately, the Horn and Schunck algorithm, for instance, or least squares in linear space parametrization [44, 60] or, yet, by over-parametrization for added accuracy and motion edge definition [44] (Sect. 3.8).

The following shows how active contours can be used to formulate joint motion estimation and segmentation. The formulation has been investigated in [54] for an arbitrarily fixed number of regions using the piecewise constant model of motion, i.e., optical flow in each segmentation region is considered constant. It has been extended to higher order linear models of motion, the affine for instance, and to the spatiotemporal domain in [103]. The expansion of motion in a general linear space of functions was studied in [60]. To bring out the main concepts involved in concurrent optical flow estimation and segmentation it is sufficient to use the piecewise constant model of motion and the case of a segmentation into two regions. We will use the method of Cremers [54] for this purpose.

Two-region partitioning.
Consider the case of segmenting the flow field into two regions and let $\mathscr{R} = \{R_i\}_1^2$ be any two-region partition of the image sequence domain Ω. Let γ be a regular closed plane curve parametrized by arc length, $\gamma : [0, l] \leftarrow \mathbb{R}^2$, where l is the curve length, such that γ and all its derivatives agree at the endpoints 0 and l. We will

further request that γ be a simple curve, i.e., that it has no other self-intersections but at $0, l$, i.e., $s_1, s_2 \in]0, l[, s_1 \neq s_2 \implies \gamma(t_1) \neq \gamma(t_2)$. Let R_γ be the interior of γ. The regions R_1 and R_2 of the two-region partition $\mathscr{R}$ of Ω will be represented by R_γ and R_γ^c, respectively. i.e., $R_1 = R_\gamma$ and $R_2 = R_\gamma^c$.

Under the piecewise constant model of optical flow, i.e, where the flow is assumed constant, equal to some velocity vector (a_i, b_i) in each region R_i of $\mathscr{R}$, the worth, or quality, of $\mathscr{R}$ as an optical flow based segmentation of the image sequence I at some time of observation can be represented by the following functional:

$$\mathscr{E}(\mathscr{R}, \{a_i, b_i\}_1^2) = \mathscr{E}(\gamma, \{a_i, b_i\}_1^2) = \sum_{i=1}^{2} \int_{R_i} e_i(x, y)dxdy + \lambda \int_\gamma ds, \tag{3.93}$$

where, for $i = 1, 2$, e_i is a function which evaluates how well the piecewise constant representation of optical flow fits the observed data, namely the spatiotemporal variations of the image sequence within R_i. For instance, we can use the squared piecewise constant parametric expression of the lefthand side of the Horn and Schunck equation, a special case of the more general representation in Eqs. 3.74–3.77. An alternative is to use the squared cosine of the angle between the image spatiotemporal gradient and the spatiotemporal velocity vector $(u, v, 1)$, i.e., the square of the dot product of the unit image spatiotemporal gradient and unit spatiotemporal velocity vector, which is just what the lefthand side of the Horn and Schunck equation expresses would we normalize the two vectors by their respective length. Precisely, if the constant velocity vector of region R_i is $\mathbf{w}_i = (a_i, b_i, 1)^T$, $i = 1, 2$, then:

$$e_i = \frac{(\mathbf{w}_i^T \nabla_3 I)^2}{\|\mathbf{w}_i\|^2 \|\nabla_3 I\|^2}, \tag{3.94}$$

where ∇_3 designates the spatiotemporal gradient, $\nabla_3 I = (I_x, I_y, I_t)^T = (\nabla I, I_t)^T$. Of course, $\mathbf{w}_i \neq 0$ because the third component of the vectors is 1. To avoid zero denominators, a small quantity may be added to the image spatiotemporal gradient norm: $\|\nabla_3 I\| + \varepsilon \leftarrow \|\nabla_3 I\|$.

The main difference between the data function of Eq. 3.94 and the one used commonly in optical flow estimation, namely the squared lefthand side of the Horn and Schunck gradient equation:

$$e_i = (I_x u + I_y v + I_t)^2, \tag{3.95}$$

is the normalization of the spatiotemporal image gradient and motion vectors occurring in Eq. (3.94). The normalization to a unit vector of the image spatiotemporal gradient gives equal strength to the contribution in the objective functional from every point of Ω where this vector is not zero. This is not the case with Eq. (3.95) which gives more weight, therefore more importance, to high contrast points. It is unclear whether high image contrast should or should not be given priority in determining optical flow. However, Eq. 3.94 has the merit, as we will see, of leading directly to the expression of a small-matrix eigenvalue problem when minimizing the objective functional with respect to the optical flow model parameters $a_i, b_i, \ i = 1, 2$.

The integral in the second term of the objective functional Eq. (3.93) is the length of γ and has the effect of shortening it, therefore smoothing it. We know from Chap. 2 that this smoothing manifests as curvature in the Euler-Lagrange equation corresponding to this term in the minimization of Eq. (3.93) with respect to γ.

Minimization with respect to the motion parameters.
Let **S** be the 3×3 matrix defined by:

$$\mathbf{S} = \frac{\nabla_3 I (\nabla_3 I)^T}{\|\nabla_3 I\|^2}. \tag{3.96}$$

This matrix is, of course, a function of image position: $\mathbf{S} = \mathbf{S}(x, y)$. The data function for each region $R_i, \ i = 1, 2$ can then be rewritten as:

$$e_i = \frac{\mathbf{w}_i^T \mathbf{S} \mathbf{w}_i}{\|\mathbf{w}_i\|^2}, \tag{3.97}$$

With this notation, differentiation with respect to $\{a_i, b_i\}_1^2$ of Eq. (3.93) under the integral sign gives for each region R_i the solution $\tilde{\mathbf{w}}_i$ defined by:

$$\tilde{\mathbf{w}}_i = \arg\min_{\mathbf{w}} \frac{\mathbf{w}^T \mathbf{M}_i \mathbf{w}}{\mathbf{w}^t \mathbf{w}}, \tag{3.98}$$

where $\mathbf{M}_i$ is the 3×3 data matrix given by:

$$\mathbf{M}_i = \int_{R_i} \mathbf{S}(x, y) dx dy, \tag{3.99}$$

obtained by integrating each element of **S** over R_i. Because $\mathbf{M}_i$ is a symmetric matrix, its smallest eigenvalue μ_i is characterized by [104]:

$$\mu_i = \min_{\mathbf{w}} \frac{\mathbf{w}^T \mathbf{M}_i \mathbf{w}}{\mathbf{w}^t \mathbf{w}}. \tag{3.100}$$

Therefore, the solution $\tilde{\mathbf{w}}_i$ is the eigenvector corresponding to this smallest eigenvalue and which has the third component equal to 1.

Minimization with respect to γ: curve evolution equation.
With the motion parameters fixed, i.e., assuming they are independent of γ (or R_γ), the functional derivative of the integral on R_γ of the objective functional data term is (see Chap. 2 for basic formulas):

$$\frac{\partial}{\partial \gamma} \int_{R_\gamma} e_1(x, y) dx dy = e_1 \mathbf{n}, \tag{3.101}$$

where $\mathbf{n}$ is the outward unit normal function of γ. Similarly for the integral over R_γ^c:

$$\frac{\partial}{\partial \gamma} \int_{R_\gamma^c} e_2(x, y) dx dy = -e_2 \mathbf{n}. \tag{3.102}$$

The minus sign on the right-hand side of Eq. (3.102) is due to the fact that the boundary of R_γ^c is $-\mathbf{n}$. The functional derivative of the length integral of Eq. (3.93) is (see Chap. 2):

$$\frac{\partial}{\partial \gamma} \int_\gamma ds = \kappa \mathbf{n}, \tag{3.103}$$

where κ is the curvature function of γ. Accounting for all its terms, the functional derivative of the objective functional Eq. (3.93) is:

$$\frac{\partial \mathscr{E}}{\partial \gamma} = (e_1 - e_2 + \lambda \kappa) \mathbf{n}, \tag{3.104}$$

Let γ be embedded in a one-parameter family of curves $\gamma(s, \tau)$ indexed by algorithmic time τ. The evolution equation to minimize the objective functional with respect to γ is (see Chap 2):

$$\frac{\partial \gamma}{\partial \tau} = -\frac{\partial \mathscr{E}}{\partial \gamma} = -(e_1 - e_2 + \lambda \kappa) \mathbf{n}, \tag{3.105}$$

Recall that the evolving curve is called an *active curve*, or an active contour.

Level set representation and evolution equation.

We recall from Chap. 2 some basic facts about level sets: an implementation of Eq. (3.105) which would explicitly discretize γ as a set of particles and move these, would be tantamount to numerical breakdown because changes in the curve topology would not be resolvable in general. Fans, where the particles separate widely to create large gaps, and shocks, where the particles come so close together as to collide or cross paths, would also be major hurdles. The level set method [105] avoids these serious problems by representing γ implicitly as a level set, the zero level set, for instance, of a function ϕ defined on the plane. The level set function ϕ is then evolved, rather than evolving γ, in a manner that is consistent with the evolution of γ, enabling the recovery of the curve at any time as its zero level set. With this representation, γ remains a well defined curve in the face of topology changes, fans, and shocks. Refer to Chap. 2 for a review.

Let the evolving curve $\gamma(s, \tau)$ be represented at all times τ by the zero level-set of function $\phi : \mathbb{R}^2 \times \mathbb{R} \to \mathbb{R}$, taken by convention to be positive inside R_γ and negative outside. The evolution equation of ϕ is given by:

$$\frac{\partial \phi}{\partial \tau} = (e_1 - e_2 + \lambda \kappa) \, \|\nabla \phi\| \tag{3.106}$$

In terms of the level set function, curvature κ of γ is expressed as:

$$\kappa = \operatorname{div}\left(\frac{\nabla\phi}{\|\nabla\phi\|}\right) \tag{3.107}$$

when the normal unit vector $\mathbf{n}$ is oriented outward:

$$\mathbf{n} = -\frac{\nabla\phi}{\|\nabla\phi\|} \tag{3.108}$$

General linear models of motion.
The method [54] is easily extended to general linear models of optical flow by writing the spatiotemporal motion vector $\mathbf{w} = (u, v, 1)^T$ in terms of the motion parameters via a transformation matrix $\mathbf{T} = \mathbf{T}(x, y)$ independent of the image. The temporal dimension can also be included in the writing. For a model of n parameters $a_1, \ldots, a_n$,

$$\mathbf{w} = \mathbf{T}\alpha, \tag{3.109}$$

where α is the vector of parameters augmented by a last element equal to 1, $\alpha = (a_1, \ldots, a_n, 1)^T$. The first half of the parameters correspond to the component u of optical flow and the other half to the v component. Matrix $\mathbf{T}$ is of size $3 \times (n+1)$. For instance, for the affine model, we have:

$$\mathbf{T} = \begin{pmatrix} x & y & 1 & 0 & 0 & 0 & 0 \\ 0 & 0 & 0 & x & y & 1 & 0 \\ 0 & 0 & 0 & 0 & 0 & 0 & 1 \end{pmatrix} \tag{3.110}$$

With this model of motion, the data function of region R_i in the objective functional becomes:

$$e_i = \frac{(\alpha_i^T \mathbf{T}\nabla_3 I)^2}{\|\alpha_i\|^2 \|\mathbf{T}\nabla_3 I\|^2}, \tag{3.111}$$

From here on, the problem statement remains the same as with the piecewise constant model of motion.

A formulation using the standard data function Eq. (3.95), rather than Eq. (3.111), and an expansion of each component of optical flow in the span of a general basis of functions as in Eqs. 3.74–3.76 of Sect. 3.8, leads to similar computations, namely an algorithm which iterates two steps, least-squares estimation of the motion parameters, which can be done efficiently by the singular value decomposition (SVD) method, and active curve evolution with a velocity of the same expression as Eq. (3.105). More specifically, and using the notation and definitions in Eqs. 3.74–3.76 of Sect. 3.8, the formulation would seek to minimize:

$$\mathscr{E}(\gamma, \alpha_1, \alpha_2) = \sum_{i=1}^{2} \int_{\mathbf{R}_i} \left(\nabla I \cdot \alpha_i^T \theta + I_t\right)^2 dxdy + \lambda \int_\gamma ds, \tag{3.112}$$

where $\alpha_i,\ i=1,2$ are the coefficient vectors for $R_i,\ i=1,2$, with $R_1=R_\gamma,\ R_2=R_\gamma^c$, and θ is the vector of basis functions. The minimization is done by iterations of two steps, one to compute the parameters by least squares (Eq. 3.78), via SVD for instance, in each region separately, and the other to evolve the curve with:

$$\frac{d\gamma}{d\tau}=-\left(\left(\nabla I\cdot\alpha_1^T\theta+I_t\right)^2-\left(\nabla I\cdot\alpha_2^T\theta+I_t\right)^2+\lambda\kappa\right)\mathbf{n}. \tag{3.113}$$

The dependence of the parameters on the segmentation does not produce extra terms [106] in the evolution equation and the minimization corresponds to gradient descent rather than simply greedy descent.

An important question arises in parametric motion estimation as to which model complexity to use, i.e., how many basis functions to use in the representation of the components of optical flow. In what concerns estimation on a given support, the higher the model order the higher the accuracy. However, when the emphasis is on segmentation, then estimation accuracy is of secondary concern as long as it is sufficient to serve the segmentation, i.e., the model order to use is the least complex that permits a distinction between the regions of segmentation. For flow fields caused by moving rigid environmental objects, or by camera motion, a low-order model such as piecewise constant or affine will probably be satisfactory. However, other flow fields, such as those due to articulated objects or elastic environmental motion, may require higher order models. Ideally, one should use the smallest order that allows discriminating between the different relevant motions of the flow field because models of higher order might represent flow variations so fine as to produce a segmentation that is an artifact of the model rather that coherent motion. This is the problem of over-fitting mentioned in [103] which observed in practice cases of reduced curve evolution stability with increased model complexity. At any rate, accurate region-confined flow estimation can always follow a reasonably correct segmentation obtained with a lower order model.

Multiregion segmentation.

A segmentation into more than two regions, called *multiregion segmentation*, or *multiphase segmentation*, uses two or more active curves. In essence, the objective functional data term for N regions $\{R_i\}$ is:

$$\mathcal{D}=\sum_{i=1}^{N}\int_{R_i}e_i(x,y)dxdy \tag{3.114}$$

If one has several active curves and uses the interior of each to define a region, one must make sure that at algorithm completion the regions so defined form a partition, i.e., that they cover the image domain and do not intersect. Therefore, one cannot simply generalize a two-region algorithm by using more curves and assigning a region to the interior of each.

Multiregion segmentation has been addressed in several different ways. The methods have been described in detail in their original papers and have also been reviewed in [31]. We will merely give of them here a brief account for a quick introduction to the literature on the subject. Matlab code of several algorithms is freely available on the web at *mathworks.de/matlabcentral/fileexchange/29447-multiphase-level-set-image-segmentation* and elsewhere.

The earliest investigations of multiregion segmentation [89, 107] addressed the problem in two quite different ways. In a region competition framework [89], the curves $\{\gamma_i\}$ were taken together as a set formed by their union, started as a partition, and moved as a set, i.e., the curve evolution equations resulting from region competition were applied to $\Gamma = \cup\gamma_i$ considered a single curve. This representation does not extent to the level set method and, as a result, Γ is tracked by discretization particles, predisposing the scheme to numerical ills which do not occur with the level set method. Along a different vein in [107], several active curves mediate multiregion segmentation, each with its own speed function, but also with a contribution from a term in the objective functional dedicated to bias the segmentation toward a partition. However, a partition is not guaranteed at algorithm convergence because this term is weighed against the others in the functional and, therefore, the weight value conditions the outcome. The scheme also appears in the investigations of image segmentation of [108, 109].

Using several active contours and stating segmentation as spatially regularized clustering, the investigations in [60, 110, 111] were able to obtain coupled functionals, one for each curve. The resulting movement of the curves ends in a partition when the curves are started so as to define a partition [110]. However, the scheme can be quite slow because it sweeps through the image several times at each of many iterations and can sometimes produce artifacts such as elongated portions along region borders.

A definite means of ensuring a partition in multiregion segmentation is simply to define a general mapping between the regions $\{R_i\}$ of the segmentation formulation and partition-defining regions drawn from the various sets which regions $\{R_{\gamma_i}\}$ form when they intersect [112, 113]. Two such mappings are shown in Fig. 3.9. Both methods guarantee a partition by construction but the computational load can quickly become excessive when the number of regions increases.

A first order-order analysis of the region data functions in the two-region case brings out the interpretation of curve evolution as point membership operations. This directs to enforcing a simple partition constraint in the multiregion case directly in the functional minimization process and which states that a point relinquished by a region is claimed by another without transition through intermediate regions, thereby maintaining implicitly a partition of the image domain at all times when segmentation is started with a partition [114, 115]. This can lead in general to very efficient execution.

Multiregion segmentation raises the question as to what the number of regions is. In general, this is just fixed to equal the number one expects to occur, but there are many cases where this is not applicable. With curve evolution methods, there have been some efforts at determining the number of regions automatically, either as part of curve evolution optimization [116] or by an external process [89, 117, 111]. However, experimentation regarding determining the number of regions automatically remains by and large insufficient, even though the question is quite important.

Example: This example (courtesy of D. Cremers) illustrates joint parametric estimation and segmentation of optical flow by the general active curve framework in [54] which we have just described. The scene of this sequence contains three circular objects moving against a mobile background. The purpose, therefore, is to segment the image into four regions on the basis of the direction of motion to be simultaneously estimated. The true image movements in the scene are: down (top left object), up (top right object), right (lower object), and left (background). The multiple region representation in terms of active contours is that of Chan and Vese [112]; therefore, two curves are needed (refer to Fig. 3.9a). The initial position of these curves is shown in Fig. 3.10a which also displays the evolving motion field. Intermediate motion fields and positions of the curves are shown in Figs. 3.10b and c. The curves and the motion field at convergence are in Fig. 3.10d. The curves define regions which correspond accurately to the objects and background, and the motion field fits the ground truth.

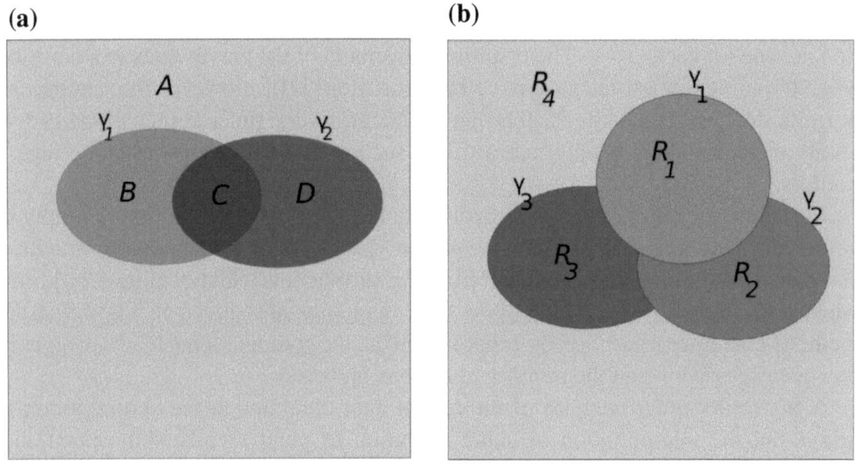

Fig. 3.9 **a** Partition construction in [112]: A single curve defines two regions. Two intersecting simple closed curves give four disjoint subsets A, B, C, D. These can be combined to have partitions of up to four regions. For four regions, $R_1 = B = R_{\gamma_1} \cap R^c_{\gamma_2}$; $R_2 = D = R_{\gamma_2} \cap R^c_{\gamma_1}$; $R_3 = C = R_{\gamma_1} \cap R_{\gamma_2}$; $R_4 = A = (R_1 \cup R_2)^c$. In general, N regions necessitate $\lceil \log N \rceil$ curves; **b** The mapping of [113]: three curves map to four partition regions: $R_1 = R_{\gamma_1}$; $R_2 = R_{\gamma_2} \cap R^c_{\gamma_1}$; $R_3 = R_{\gamma_3} \cap R^c_{\gamma_2} \cap R^c_{\gamma_1}$; $R_4 = \left(\cup_{i=1}^3 R_i\right)^c$. In general, the mapping requires $N-1$ curves for N regions

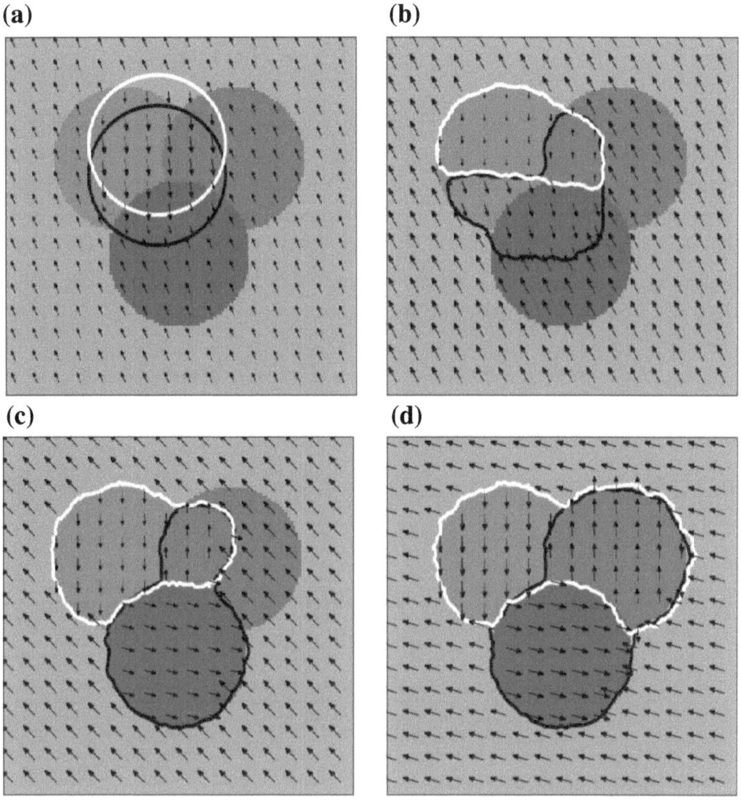

Fig. 3.10 Joint segmentation and parametric estimation of optical flow by the Cremers method (Courtesy of Daniel Cremers): The true motions in the scene are: vertically down for the top left object, vertically up for the top right object, horizontally to the right for the lower object, and horizontally to the left for the background. Two curves are used to represent four regions according to the Chan and Vese mapping (Fig. 3.9a). The initial position of the curves is shown in **a**, intermediate positions and the evolving motion field are displayed in **b** and **c**. The final segmentation and motion field are shown in **d**. Both the segmentation and the motion field fit the ground truth

3.12 Joint Optical Flow and Disparity Estimation

In stereoscopy, the disparity field and optical flow are related by the *stereokinematic constraint* [62, 63]. Therefore, their joint estimation, via this constraint, can be advantageous [118–124]. Joint estimation involves computing two motion fields and two disparity fields using the stereokinematic constraint [62, 63] which relates three of these fields to the fourth.

Let the image sequence be a real positive function over a domain $\Omega \times]0, S[\times]0, T[$, where $]0, T[$ is an interval of time, and $]0, S[$ an interval of $\mathbb{R}$:

$$I : \Omega \times]0, S[\times]0, T[\mapsto \mathbb{R}$$
$$\mathbf{x}, s, t \mapsto I(\mathbf{x}, s, t)$$

Variable s can be thought of as the parameter of the trajectory of a sequence of image planes in these planes coordinate domain. For a fixed value of s we have a temporal image sequence of images and for two distinct fixed values we obtain a stereoscopic image sequence. Therefore, this generalizes the definition of a stereoscopic image sequence. Let $(\mathbf{x}, s, t) \in \Omega \times]0, S[\times]0, T[$ and $\mathbf{x} + \mathbf{d}(\mathbf{x}, s + ds, t + dt)$ its correspondent at $(s+ds, t+dt)$, where $\mathbf{d}$ designates a displacement. The assumption that I does not change at corresponding points,

$$I(\mathbf{x} + \mathbf{d}(\mathbf{x}, s + ds, t + dt), s + ds, t + dt) = I(\mathbf{x}, s, t)$$

gives the following motion and disparity constraints:

$$\begin{aligned} \nabla I \cdot W + I_t &= 0 \\ \nabla I \cdot D + I_s &= 0, \end{aligned} \tag{3.115}$$

where W is the optical velocity vector, D the disparity vector, ∇I the spatial gradient of I, I_t and I_s the partial derivatives of I with respect to t and s. Because

$$W = \frac{\partial \mathbf{d}}{\partial t}, \qquad D = \frac{\partial \mathbf{d}}{\partial s}, \tag{3.116}$$

we also have the integrability constraint:

$$\frac{\partial W}{\partial s} = \frac{\partial D}{\partial t} \tag{3.117}$$

The integrability constraint is the continuous-disparity form of the stereokinematic constraint of [62] which was written for optical flow in discrete-disparity stereoscopic image sequences.

A fully discrete version of the integrability constraint can be written as follows. Let $I^{l,t}$, $I^{r,t}$ be the left and right images at time t and $I^{l,t'}$, $I^{r,t'}$ the left and right images at the next time t'. Let $W^{l,t} = (u^{l,t}, v^{l,t})$ and $W^{r,t} = (u^{r,t}, v^{r,t})$ designate left and right optical motion vectors at time t, and $D^t = (\delta_1^t, \delta_2^t)$, $D^{t'} = (\delta_1^{t'}, \delta_2^{t'})$ the disparity vectors at t and t'. A discrete representation of the integrability constraint can then be written:

$$W^{r,t} - W^{l,t} = D^{t'} - D^t \tag{3.118}$$

This is the *quadrilateral expression* of the stereokinematic constraint (Fig. 3.11). It is the expression generally used in practice [118–122].

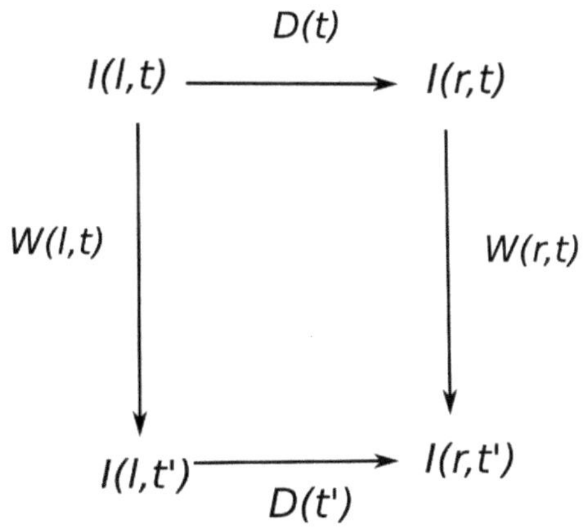

Fig. 3.11 Quadrilateral representing the stereokinematic constraint Eq. 3.118: Knowing three sides, we can deduce the fourth.

The two motion fields and the two disparity fields are related via the stereokinematic constraint. There is no other relation between any three of them but through the fourth. Therefore, joint estimation of the four fields which would treat the left and right data of stereoscopy even-handedly, can proceed along the following two veins:

(1) The four fields are estimated concurrently, for instance by minimizing an objective functional containing data and smoothness terms for each field, and a term to account for the stereokinematic constraint to bind the fields together. A slightly more efficient version of this, computationally, would estimate concurrently three of the fields bound by the stereokinematic constraint which uses the fourth field computed beforehand independently.

(2) Three of the fields are estimated independently, for instance by a variational formulation such as we have seen, and the fourth field is deduced directly using the stereokinematic constraint.

Estimation along the first vein entails solving a very large system of equations, nonlinear equations when using discontinuity preserving formulations. For instance, with a 400×400 image, the number of scalar variables to determine in four fields is $2^8 \times 10^4$ at each instant of observation. Along the second vein, one would would solve independently three much smaller problems to estimate three fields, and follow with a direct application of the stereokinematic constraint to compute the fourth field. The process, therefore, is much more efficient along this vein. Also, and as we shall see, prolonging the estimation through time can be done at each instant of time by computing independently only the two flow fields, followed by an execution of the stereokinematic constraint using the previously computed disparity field.

According to the second paradigm, whereby three fields are computed independently and the fourth deduced, we can use constraints Eq. (3.115) to estimate separately the left and right motion fields and the disparity field at time t before computing the disparity field at time t' using the integrability/stereokinematic constraint Eq. (3.118). The left and right motion fields at t can be estimated for instance as in Section 3.5 by solving:

Table 3.2 The two lines in the top box show the ground truth (constant) disparity for the background (B) and each of the two objects (O_1 and O_2) between the left and right images of the second stereoscopic image (Fig. 3.12a,b)

Actual	B	O_1	O_2
x	−1.0	1.0	1.0
y	0.0	1.0	1.0
Joint estimation	B	O_1	O_2
x	−0.8	0.96	0.92
y	0.02	0.81	0.87
Deriche et al.	B	O_1	O_2
x	−0.81	0.87	0.90
y	0.02	0.78	0.87

The middle box displays the average disparities, for the background and each object, computed by joint estimation optical flow and disparity as described in the text. The bottom box gives the average disparities computed by the Deriche-Aubert-Kornprobst method

$$W^{\{l,r\},t} = \arg\min_{W}\{\int_{\Omega}\Big((\nabla I^{\{l,r\},t}\cdot W + I_t^{\{l,r\},t})^2 + \lambda(g(\|\nabla u\|) + g(\|\nabla v\|))\Big)\,d\mathbf{x} \tag{3.119}$$

The disparity field can be computed by variational methods in a similar fashion, with or without the epipolar constraint [87, 125–128].

When the left and right motion fields and the disparity field are estimated at time t, the disparity field at time t' is deduced using the integrability/stereokinematic constraint, i.e.,

$$D^{t'} = W^{r,t} - W^{l,t} + D^t \tag{3.120}$$

We can make two observations: (1) Initially, the disparity field at time t is computed independently. Subsequently, the current disparity field is the disparity field computed at the previous instant, i.e., at each instant of time, except at the start, only the two motion fields are computed by Eq. (3.119), followed by an application of Eq. 3.120 and, (2) the formulation assumes that the disparity and motion are both of small extent. In the presence of either motion or disparity of large extent, estimation must resort to some form of multiresolution/multigrid computations (Sect. 3.10).

Example: The second of the two stereoscopic pairs of images used (constructed from the *Aqua* sequence) in this verification example (from [123]) is displayed in Fig. 3.12a, b. The scene consists of a circular object on the left (sea shell like) and a circular object on its right (sponge like), against a background (in an aquarium). Both objects are cutouts from real images.The background and the objects are given disparities in the first stereoscopic pair of images and are made to move such that disparities in the second stereoscopic pair are (−1,0) for the background, and (1,1) for the two objects (Table 3.2 upper box). The results are shown graphically in Fig. 3.12 for a qualitative appraisal, and quantitatively in Table 3.2 (lower two boxes).

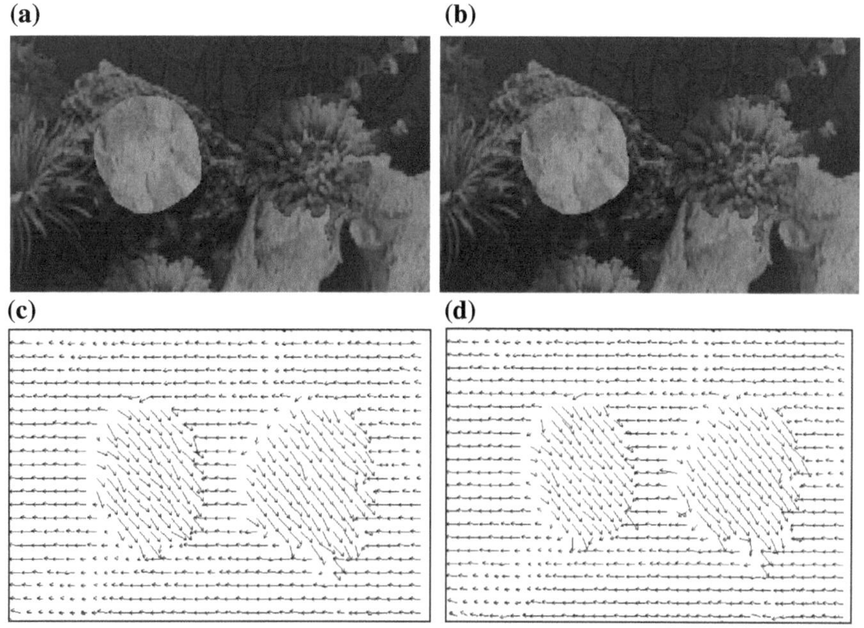

Fig. 3.12 Joint estimation of small extent optical flow and disparity: **a, b** the second of the two pairs of stereoscopic images used; **c** a graphical display of the disparities computed by joint estimation using the integrability/stereokinematic constraint; **d** A graphical display of the disparities computed with the Deriche-Aubert-Kornprobst method

3.13 State-of-the-Art

This chapter has presented the fundamental concepts underlying optical flow and its estimation, namely (i) the optical flow constraint which relates optical velocity to the image spatiotemporal gradient, (ii) the variational principle and the basic roles that conformity to data and regularization play in problem formulations, (iii) the necessity and mechanisms to preserve the sharpness of motion boundaries, (iv) mutiresolution/multigrid processing to deal with long-range motion, (v) the combination of motion segmentation and motion estimation as joint processes, and (vi) the concurrent estimation of the optical flow and disparity fields. These concepts are self-contained and, as such, they were described separately to allow their full meaning to be exposed unconcealed by other considerations. Algorithms which account for each concept have been described, such as the Horn and Schunck method, the Deriche-Aubert-Kornprobst's and the Cremers'. The purpose of the presentation was to focus on explaining the idea underlying each abstraction and on means of effecting it, and no attempt was made to describe algorithms that would embody together several concepts. Such algorithms have been the concern of a number of studies investigating various mechanisms for accurate estimation. The domain is now mature enough to allow a thorough treatment of the problem leading to fast, effective, and accurate algorithms with results that can be used for a variety of useful purposes, including motion detection and three-dimensional structure and motion recovery. This is the case, for instance, with the investigations in [12, 41, 42] which describe detailed optical flow computations that have produced impressive results. Faster computations using the conjugate gradient method to solve a large linear system of equations, rather than Gauss-Seidel or similar, have been implemented in Matlab/C^{++} and made available by [129] (http://people.csail.mit.edu/celiu/OpticalFlow/). The availability of good algorithms and implementations is complemented by useful collections of test image sequences and motion data, notably the Middlebury database (http://vision.middlebury.edu/flow/). Finally, the successful computational formulations and mechanisms used in optical flow estimation have found good use in joint disparity and optical flow estimation [124].

References

1. J.J. Gibson, *The Perception of the Visual World* (Houghton Mifflin, Boston, 1950)
2. K. Nakayama, Biological image motion processing: a survey. Vision. Res. **25**, 625–660 (1985)
3. H.-H. Nagel, On the estimation of optical flow: relations between different approaches and some new results. Artif. Intell. **33**(3), 299–324 (1987)
4. H.-H. Nagel, Image sequence evaluation: 30 years and still going strong, in *International Conference on Pattern Recognition*, 2000, pp. 1149–1158
5. J. Barron, D. Fleet, S. Beauchemin, Performance of optical flow techniques. Int. J. Comput. Vision **12**(1), 43–77 (1994)
6. A. Mitiche, *Computational Analysis of Visual Motion* (Plenum Press, New York, 1994)
7. A. Mitiche, P. Bouthemy, Computation and analysis of image motion: A synopsis of current problems and methods. Int. J. Comput. Vision **19**(1), 29–55 (1996)
8. C. Stiller, J. Konrad, Estimating motion in image sequences: A tutorial on modeling and computation of 2D motion. IEEE Signal Process. Mag. **16**(4), 70–91 (1999)
9. G. Aubert, P. Kornpbrost, *Mathematical Problems in Image Processing: Partial Differential Equations and the Calculus of Variations* (Springer, New York, 2006)
10. B. Horn, B. Schunck, Determining optical flow. Artif. Intell. **17**, 185–203 (1981)
11. B.D. Lucas, T. Kanade, An iterative image registration technique with an application to stereo vision, in *IJCAI*, 1981, pp. 674–679
12. A. Bruhn, J. Weickert, C. Schnörr, Lucas/kanade meets Horn/Schunck: combining local and global optic flow methods. Int. J. Comput. Vision **61**(3), 211–231 (2005)
13. A. Mitiche, A. Mansouri, On convergence of the Horn and Schunck optical flow estimation method. IEEE Trans. Image Process. **13**(6), 848–852 (2004)
14. C. Koch, J. Luo, C. Mead, J. Hutchinson, Computing motion using resistive networks, in *NIPS*, 1987, pp. 422–431
15. J. Hutchinson, C. Koch, J. Luo, C. Mead, Compting motion using analog and binary resistive networks. IEEE Comput. **21**(3), 52–63 (1988)
16. H. Nagel, W. Enkelmann, An investigation of smoothness constraints for the estimation of displacement vector fields from image sequences. IEEE Trans. Pattern Anal. Mach. Intell. **8**(5), 565–593 (1986)
17. H.-H. Nagel, On a constraint equation for the estimation of displacement rates in image sequences. IEEE Trans. Pattern Anal. Mach. Intell. **11**(1), 13–30 (1989)
18. H.-H. Nagel, Extending the 'oriented smoothness constraint' into the temporal domain and the estimation of derivatives of optical flow, in *European Conference on Computer Vision*, 1990, pp. 139–148

19. M. Snyder, On the mathematical foundations of smoothness constraints for the determination of optical flow and for surface reconstruction. IEEE Trans. Pattern Anal. Mach. Intell. **13**(11), 1105–1114 (November 1991)
20. A. Mansouri, A. Mitiche, J. Konrad, Selective image diffusion: application to disparity estimation, in *International Conference on Image Processing*, 1998, pp. 284–288
21. F. Hampel, E. Ronchetti, P. Rousseeuw, W. Stahel, *Robust Statistics: The Approach Based on Influence Functions* (Wiley-Interscience, New York, 1986)
22. M. Black, "Robust incremental optical flow", in Ph.D. Thesis, Yale University, Research Report YALEU-DCS-RR-923, 1992
23. M. J. Black, P. Anandan, A framework for the robust estimation of optical flow, in *International Conference on Computer Vision*, 1993, pp. 231–236
24. A. Blake, A. Zisserman, *Visual Reconstruction* (MIT Press, Cambridge, 1987)
25. E. Memin, P. Perez, Joint estimation-segmentation of optic flow, in *European Conference on Computer Vison*, 1998, vol. II, pp. 563–578
26. S. Geman, D. Geman, Stochastic relaxation, gibbs distributions, and the bayesian restoration of images. IEEE Trans. Pattern Anal. Mach. Intell. **6**(6), 721–741 (1984)
27. J. Konrad, E.Dubois, Bayesian estimation of motion vector fields. IEEE Trans. Pattern Anal. Mach. Intell. **14**(9), 910–927 (1992)
28. F. Heitz, P. Bouthemy, Multimodal estimation of discontinuous optical flow using markov random fields. IEEE Trans. Pattern Anal. Mach. Intell. **15**(12), 1217–1232 (1993)
29. J. Zhang, G.G. Hanauer, The application of mean field theory to image motion estimation. IEEE Trans. Image Process. **4**(1), 19–33 (1995)
30. P. Nesi, Variational approach to optical flow estimation managing discontinuities. Image Vision Comput. **11**(7), 419–439 (1993)
31. A. Mitiche, I. Ben Ayed, *Variational and Level Set Methods in Image Segmentation* (Springer, New York, 2010)
32. D. Mumford, J. Shah, Boundary detection by using functionals. Comput. Vis. Image Underst. **90**, 19–43 (1989)
33. Y.G. Leclerc, Constructing simple stable descriptions for image partitioning. Int. J. Comput. Vision **3**(1), 73–102 (1989)
34. J. Weickert, A review of nonlinear diffusion filtering, in *Scale-Space*, 1997, pp. 3–28
35. L. Blanc-Feraud, M. Barlaud, T. Gaidon, Motion estimation involving discontinuities in a multiresolution scheme. Opt. Eng. **32**, 1475–1482 (1993)
36. M. Proesmans, L.J.V. Gool, E.J. Pauwels, A. Oosterlinck, Determination of optical flow and its discontinuities using non-linear diffusion, in *European Conference on Computer Vision*, 1994, pp. 295–304
37. R. Deriche, P. Kornprobst, G. Aubert, Optical-flow estimation while preserving its discontinuities: A variational approach, in *Asian Conference on Computer Vision*, 1995, pp. 71–80
38. G. Aubert, R. Deriche, P. Kornprobst, Computing optical flow via variational thechniques. SIAM J. Appl. Math. **60**(1), 156–182 (1999)
39. A. Kumar, A. Tannenbaum, G.J. Balas, Optical flow: a curve evolution approach. IEEE Trans. Image Process. **5**(4), 598–610 (1996)
40. J. Weickert, C. Schnörr, Variational optic flow computation with a spatio-temporal smoothness constraint. J Math. Imaging Vision **14**(3), 245–255 (2001)
41. T. Brox, A. Bruhn, N. Papenberg, J. Weickert, High accuracy optical flow estimation based on a theory for warping, 2004. http://citeseer.ist.psu.edu/brox04high.html.
42. N. Papenberg, A. Bruhn, T. Brox, S. Didas, J. Weickert, Highly accurate optic flow computation with theoretically justified warping. Int. J. Comput. Vision **67**(2), 141–158 (2006)
43. C. Zach, T. Pock, H. Bischof, A duality based approach for realtime tv-l1 optical flow, in *Annual Symposium of the German Association Pattern Recognition*, 2007, pp. 214–223
44. T. Nir, A.M. Bruckstein, R. Kimmel, Over-parameterized variational optical flow. Int. J. Comput. Vision **76**(2), 205–216 (2008)
45. M. Werlberger, W. Trobin, T. Pock, A. Wedel, D. Cremers, H. Bischof, Anisotropic huber-l1 optical flow, in *BMVC*, 2009

46. C. Vogel, *Computational Methods for Inverse Problems* (SIAM, Philadelphia, 2002)
47. L. I. Rudin, S. Osher, Total variation based image restoration with free local constraints, in *ICIP*, vol. 1, 1994, pp. 31–35
48. I. Cohen, Nonlinear variational method for optical flow computation, in *SCIA93*, 1993, pp. 523–530
49. P. Perona, J. Malik, Scale space and edge detection using anisotropic diffusion. IEEE Trans. Pattern Anal. Mach. Intell. **12**(7), 629–639 (1981)
50. G. Bellettini, On the convergence of discrete schemes for the perona-malik equation. Proc. Appl. Math. Mech.**7**(1), 1023401–1023402 (2007)
51. W. Enkelmann, Investigation of multigrid algorithms for the estimation of optical flow fields in image sequences. Computer Vision, Graphics, and Image Processing **43**(2), 150–177 (August 1988)
52. D. Terzopoulos, Efficient multiresolution algorithms for computing lightness, shape-from-shading, and optical flow, in *AAAI conference*, 1984, pp. 314–317
53. D. Cremers, C. Schnorr, Motion competition: Variational integration of motion segmentation and shape regularization, in *DAGM Symposium on, Pattern Recognition*, 2002, pp. 472–480
54. D. Cremers, A multiphase level set framework for motion segmentation, in *Scale Space Theories in Computer Vision*, ed. by L. Griffin, M. Lillholm (Springer, Isle of Skye, 2003), pp. 599–614
55. A. Mansouri, J. Konrad, Multiple motion segmentation with level sets. IEEE Trans. Image Process. **12**(2), 201–220 (Feb. 2003)
56. D. Cremers, S. Soatto, Motion competition: A variational approach to piecewise parametric motion segmentation. Int. J. Comput. Vision **62**(3), 249–265 (2005)
57. T. Brox, A. Bruhn, J. Weickert, Variational motion segmentation with level sets, in *European Conference on Computer Vision*, vol. 1, 2006, pp. 471–483
58. H. Sekkati, A. Mitiche, Joint optical flow estimation, segmentation, and 3D interpretation with level sets. Comput. Vis. Image Underst. **103**(2), 89–100 (2006)
59. H. Sekkati, A. Mitiche, Concurrent 3D motion segmentation and 3D interpretation of temporal sequences of monocular images. IEEE Trans. Image Process. **15**(3), 641–653 (2006)
60. C. Vazquez, A. Mitiche, R. Laganiere, Joint segmentation and parametric estimation of image motion by curve evolution and level sets. IEEE Trans. Pattern Anal. Mach. Intell. **28**(5), 782–793 (2006)
61. A. Mitiche, H. Sekkati, Optical flow 3D segmentation and interpretation: A variational method with active curve evolution and level sets. IEEE Trans. Pattern Anal. Mach. Intell. **28**(11), 1818–1829 (Nov. 2006)
62. A. Mitiche, On combining stereopsis and kineopsis for space perception, in *IEEE Conference on Artificial Intelligence Applications*, 1984, pp. 156–160
63. A. Mitiche, A computational approach to the fusion of stereopsis and kineopsis, in *Motion Understanding: Robot and Human Vision*, ed. by W.N. Martin, J.K. Aggarwal (Kluwer Academic Publishers, Boston, 1988), pp. 81–99
64. S. Negahdaripour, C. Yu, A generalized brightness change model for computing optical flow, in *ICCV*, 1993, pp. 2–11
65. M. Mattavelli, A. Nicoulin, Motion estimation relaxing the constancy brightness constraint, in *ICIP*, vol. 2, 1994, pp. 770–774
66. R.P. Wildes, M.J. Amabile, A.-M. Lanzillotto, T.-S. Leu, Physically based fluid flow recovery from image sequences, in *CVPR*, 1997, pp. 969–975
67. T. Corpetti, É. Mémin, P. Pérez, Dense estimation of fluid flows. IEEE Trans. Pattern Anal. Mach. Intell. **24**(3), 365–380 (2002)
68. K. Nakayama, S. Shimojo, Intermediate and higher order aspects of motion processing, in *Neural Mechanisms of Visual Perception*, ed. by D.M-K. Lam, C.D. Gilbert (Portfolio Publishing Company, The Woodlands, Texas, 1989), pp. 281–296
69. D. Todorovic, A gem from the past: Pleikart Stumpf's (1911) anticipation of the aperture problem, reichardt detectors, and perceived motion loss at equiluminance. Perception **25**(10), 1235–1242 (1996)

70. E. Hildreth, *The Measurement of Visual Motion* (MIT Press, Cambridge, 1983)
71. J. Kearney, W. Thompson, D. Boley, Optical flow estimation: an error analysis of gradient-based methods with local optimization. IEEE Trans. Pattern Anal. Mach. Intell. **9**(2), 229–244 (1987)
72. K. Wohn, L.S. Davis, P. Thrift, Motion estimation based on multiple local constraints and nonlinear smoothing. Pattern Recogn. **16**(6), 563–570 (1983)
73. V. Markandey, B. Flinchbaugh, Multispectral constraints for optical flow computation, in *International Conference on Computer Vision*, 1990, pp. 38–41
74. A. Mitiche, Y.F. Wang, J.K. Aggarwal, Experiments in computing optical flow with the gradient-based, multiconstraint method. Pattern Recogn. **20**(2), 173–179 (1987)
75. O. Tretiak, L. Pastor, Velocity estimation from image sequences with second order differential operators, in *International Conference on Pattern Recognition and Image Processing*, 1984, pp. 16–19
76. A. Verri, F. Girosi, V. Torre, Differential techniques for optical flow. J. Opt. Soc. Am. A **7**, 912–922 (May 1990)
77. M. Campani, A. Verri, "Computing optical flow from an overconstrained system of linear algebraic equations, in *International Conference on Computer Vision*, 1990, pp. 22–26
78. M. Tistarelli, Computation of coherent optical flow by using multiple constraints, in *International Conference on Computer Vision*, 1995, pp. 263–268
79. R. Woodham, Multiple light source optical flow, in *International Conference on Computer Vision*, 1990, pp. 42–46
80. S. Baker, I. Matthews, Lucas-kanade 20 years on: a unifying framework. Int. J. Comput. Vision **56**(3), 221–255 (2004)
81. G. Dahlquist, A. Bjork, *Numerical Methods* (Prentice Hall, Englewood Cliffs, 1974)
82. P. Ciarlet, *Introduction à l'analyse numérique matricielle et à l'optimisation*, 5th edn. (Masson, Paris, 1994)
83. J. Stoer, P. Burlisch, *Introduction to Numerical Methods*, 2nd edn. (Springer, New York, 1993)
84. R. Feghali, A. Mitiche, Fast computation of a boundary preserving estimate of optical flow. SME Vision Q. **17**(3), 1–4 (2001)
85. L. Yuan, J. Li, B. Zhu, Y. Qian, A discontinuity-preserving optical flow algorithm, in *IEEE International Symposium on Systems and Control in Aerospace and Aeronautics*, 2006, pp. 450–455
86. W. Enkelmann, K. Kories, H.-H. Nagel, G. Zimmermann, An experimental investigation of estimation approaches for optical flow fields, in *Motion Understanding: Robot and Human Vision*, ed. by W.N. Martin, J.K. Aggarwal, (Chapter 6), (Kluwer Academic Publications, Boston, 1988), pp. 189–226
87. L. Álvarez, J. Weickert, J. Sánchez, Reliable estimation of dense optical flow fields with large displacements. Int. J. Comput. Vision **39**(1), 41–56 (2000)
88. S. Solimini, J.M. Morel, *Variational Methods in Image Segmentation* (Springer, New York, 2003)
89. S. Zhu, A. Yuille, Region competition: Unifying snakes, region growing, and bayes/mdl for multiband image segmentation. IEEE Trans. Pattern Anal. Mach. Intell. **118**(9), 884–900 (1996)
90. T. Brox, B. Rosenhahn, D. Cremers, H.-P. Seidel, High accuracy optical flow serves 3- D pose tracking: Exploiting contour and flow based constraints, in *European Conference on Computer Vision*, 2006, pp. 98–111
91. C. Zach, T. Pock, H. Bischof, A duality based approach for realtime tv-l1 optical flow, in *DAGM*, 2007, pp. 214–223
92. A. Wedel, T. Pock, C. Zach, H. Bischof, D. Cremers, An improved algorithm for tv-l1 optical flow, in *Statistical and Geometrical Approaches to Visual Motion Analysis*, ed. by D. Cremers, B. Rosenhahn, A. Yuille, F. Schmidt. ser. Lecture Notes in Computer Science, (Springer, Heidelberg, 2009), pp. 23–45
93. A. Foi, M. Trimeche, V. Katkovnik, K. Egiazarian, Practical poissonian-gaussian noise modeling and fitting for single-image raw-data. IEEE Trans. Image Process. **17**(10), 1737–1754 (2008)

94. J. Bergen, P. Anandan, K. Hanna, R. Hingorani, Hierarchical model-based motion estimation, in *European Conference on Computer Vision*, 1992, pp. 237–252
95. É. Mémin, P. Pérez, Dense estimation and object-based segmentation of the optical flow with robust techniques. IEEE Trans. Image Process. **7**(5), 703–719 (1998)
96. P. Burt, E. Adelson, The Laplacian pyramid as a compact image code. IEEE Trans. Commun. **31**(4), 532–540 (April 1983)
97. F. Heitz, P. Perez, P. Bouthemy, Multiscale minimization of global energy functions in some visual recovery problems. CVGIP: Image Underst. **59**(1), 125–134 (1994)
98. C. Cassisa, V. Prinet, L. Shao, S. Simoens, C.-L. Liu, Optical flow robust estimation in a hybrid multi-resolution mrf framework, in *IEEE Acoustics, Speech, and, Signal Processing*, 2008, pp. 793–796
99. W. Hackbusch, U.Trottenberg (eds.), *Multigrid Methods. Lecture Notes in Mathematics*, vol. 960, (Springer, New York, 1982)
100. W.L. Briggs, *A Multigrid Tutorial* (SIAM, Philadelphia, 1987)
101. M. Chang, A. Tekalp, M. Sezan, Simultaneous motion estimation and segmentation. IEEE Trans. Image Process. **6**(9), 1326–1333 (1997)
102. D. Cremers, A. Yuille, A generative model based approach to motion segmentation, in *German Conference on Pattern Recognition (DAGM)*, (Magdeburg, Sept 2003), pp. 313–320
103. D. Cremers, S. Soatto, Variational space-time motion segmentation, in *International Conference on Computer Vision*, vol 2 (Nice, France, 2003), pp. 886–892
104. R.A. Horn, C.R. Johnson, *Matrix Analysis* (Cambridge University Press, Cambridge, 1985)
105. J.A. Sethian, *Level Set Methods and Fast Marching Methods* (Cambridge University Press, Cambridge, 1999)
106. G. Aubert, M. Barlaud, O. Faugeras, S. Jehan-Besson, Image segmentation using active contours: calculus of variations or shape gradients? SIAM J. Appl. Math. **63**(6), 2128–2154 (2003)
107. H.-K. Zhao, T. Chan, B. Merriman, S. Osher, A variational level set approach to multiphase motion. J. Comput. Phys. **127**(1), 179–195 (1996)
108. N. Paragios, R. Deriche, Coupled geodesic active regions for image segmentation: A level set approach, in *Europeean Conference on Computer vision*, (Dublin, Ireland, June 2000), pp. 224–240
109. C. Samson, L. Blanc-Feraud, G. Aubert, J. Zerubia, A level set model for image classification. Int. J. Comput. Vision **40**(3), 187–197 (2000)
110. C. Vazquez, A. Mitiche, I. Ben Ayed, Image segmentation as regularized clustering: A fully global curve evolution method, in *International Conference on Image Processing*, 2004, pp. 3467–3470
111. T. Brox, J. Weickert, Level set segmentation with multiple regions. IEEE Trans. Image Process. **15**(10), 3213–3218 (2006)
112. L. Vese, T. Chan, A multiphase level set framework for image segmentation using the Mumford and Shah model. Int. J. Comput. Vision **50**(3), 271–293 (2002)
113. A. Mansouri, A. Mitiche, C. Vazquez, Multiregion competition: a level set extension of region competition to multiple region partioning. Comput. Vis. Image Underst. **101**(3), 137–150 (2006)
114. I. Ben Ayed, A. Mitiche, Z. Belhadj, Polarimetric image segmentation via maximum likelihood approximation and efficient multiphase level sets. IEEE Trans. Pattern Anal. Mach. Intell. **28**(9), 1493–1500 (2006)
115. I. Ben Ayed, A. Mitiche, A partition constrained minimization scheme for efficient multiphase level set image segmentation, in *International Conference on Image Processing*, 2006, pp. 1641–1644
116. I. Ben Ayed, A. Mitiche, A region merging prior for variational level set image segmentation. IEEE Trans. Image Process. **17**(12), 2301–2313 (2008)
117. T. Kadir, M. Brady, Unsupervised non-parametric region segmentation using level sets, in *International Conference on Computer Vision*, 2003, pp. 1267–1274

118. A. Tamtaoui, C. Labit, Constrained disparity and motion estimators for 3DTV image sequence coding. Signal Proces.: Image Commun. **4**(1), 45–54 (1991)
119. J. Liu, R. Skerjanc, Stereo and motion correspondence in a sequence of stereo images. Signal Process.: Image Commun. **5**(4), 305–318 (October 1993)
120. Y. Altunbasak, A. Tekalp, G. Bozdagi, Simultaneous motion-disparity estimation and segmentation from stereo, in *IEEE International Conference on Image Processing*, vol. III, 1994, pp. 73–77
121. R. Laganière, Analyse stéreocinétique d'une séquence d'images: Estimation des champs de mouvement et de disparité, Ph.D. dissertation, Institut national de la recherche scientifique, INRS-EMT, 1995
122. I. Patras, N. Alvertos, G. Tziritas, Joint disparity and motion field estimation in stereoscopic image sequences, in *IAPR International Conference on Pattern Recognition*, vol. I, 1996, pp. 359–363
123. H. Weiler, A. Mitiche, A. Mansouri, Boundary preserving joint estimation of optical flow and disparity in a sequence of stereoscopic images, in *International Conference on Visualization, Imaging, and Image Processing*, 2003, pp. 102–106
124. A. Wedel, T. Brox, T. Vaudrey, C. Rabe, U. Franke, D. Cremers, Stereoscopic scene flow computation for 3D motion understanding. Int. J. Comput. Vision **95**(1), 29–51 (2011)
125. L. Robert, R. Deriche, Dense depth map reconstruction: A minimization and regularization approach which preserves discontinuities, in *European Conference on Computer Vision*, 1996, pp. I:439–451
126. O. Faugeras, R. Keriven, Variational principles, surface evolution, PDEs, level set methods, and the stereo problem. IEEE Trans. Image Process. **7**(3), 336–344 (1998)
127. H. Zimmer, A. Bruhn, L. Valgaerts, M. Breuß, J. Weickert, B. Rosenhahn, H.-P. Seidel, PDE-based anisotropic disparity-driven stereo vision, in *Vision Modeling and Visualization*, 2008, pp. 263–272
128. C. Wohler, *3D Computer Vision: Efficient Methods and Applications* (Springer, Berlin, 2009)
129. C. Liu, Beyond pixels: Exploring new representations and applications for motion analysis, in *Ph.D. Thesis*, MIT, May 2009

Chapter 4
Motion Detection

4.1 Introduction

In the context of motion analysis, the image *foreground* is the region of the image domain which corresponds to the projected surfaces of the moving environmental objects, and the *background* is its complement. *Motion detection* separates the domain of an image sequence into foreground and background. Because it refers to environmental object motion, this general definition is valid for both static and moving viewing systems. When the viewing system is static, the foreground motion is exclusively due to the projected surfaces of the objects in motion. When the viewing system moves, it causes image motion which combines by vector addition with the image motion due to object movement. In this case, foreground detection requires that the motion due to the viewing system movement be accounted for, for instance by subtracting it from the combined image motion so that the residual motion is due to the moving objects only.

Motion detection does not necessarily have to evaluate image motion, unlike *motion segmentation* which divides the image domain into regions corresponding to distinct motions and which, therefore, must distinguish between the various differently moving image objects by evaluating their motion, or the motion of the environmental objects which induced it. However, image motion can be the basis of detection when available.

Detection may be done without explicit recourse to the motion field in two different ways. One scheme is to use a *background template*. The template is an image of the environment considered typical when none of the anticipated object motions occur. Motion in an image is then detected by comparing the image to the template. The other way of detecting motion without computing it first is to use the image temporal derivative, which amounts to successive frames differencing in the case of digital image sequences. In this case, the derivative is evaluated to determine whether it is due to environmental motion or not, which, ultimately, amounts to deciding whether an image change at a point is significant and attributable to object motion rather than to noise or other imaging artifact.

A. Mitiche and J. K. Aggarwal, *Computer Vision Analysis of Image Motion by Variational Methods*, Springer Topics in Signal Processing 10, DOI: 10.1007/978-3-319-00711-3_4,

Fig. 4.1 Examples of scenes in surveillance and scene monitoring applications. The intensity at any given point and time in such scenes is affected generally by sensing noise. **a** The intensity can be significantly altered by illumination variations in the course of the day. **b** The tree foliage can have local movements that produce significant variations in the sensed image intensity; yet this motion is irrelevant. **c** Snow produces sensing artifacts which can prevent the detection of moving targets

At first consideration, motion detection might be deemed an easy task because it seems to require a mere comparison of an image to a template or of an image to a subsequent one in a sequence, or an estimation of image motion and a simple analysis of it to determine where in the image domain it is significant, for instance by a thresholding operation which would declare part of the static background any point where the amount of motion is under a cutoff value. These simple ways of detecting motion are in fact commonly in use in many applications. However, as the following few examples indicate, the problem is not so simple in reality.

The illumination and sensing conditions of the building scene imaged in Fig. 4.1a (image sequence from the Video Surveillance Online Repository, VISOR, http://imagelab.ing.unimore.it/visor/video_videosInCategory.asp?idcategory=11) are common in surveillance and scene monitoring applications. The intensity at any given point and time is affected in such scenes by sensing noise and can be significantly altered by illumination variations in the course of the day. When the camera and the scene are motionless, one may think that an image sequence recorded during a short interval of time will be just about constant. However, this is not generally the case, as the graphs of Fig. 4.2a, b illustrate well. The plots show the grey level intensity in consecutive frames for two different image domain grid points, during an interval of time when both the camera and the scene are motionless. The intensity variations from a frame to the next in these plots foretell that even image differencing, the simplest method for motion detection, will require some form of noise modelling or of spatial/temporal regularization to be of any practical use.

Objects such as trees, commonly appearing in surveillance scenes, as in Fig. 4.1b (image sequence from the Video Surveillance Online Repository, VISOR, http://imagelab.ing.unimore.it/visor/video_videosInCategory.asp?idcategory=11), or Fig. 4.3, pp. 103 (courtesy of Prof. Robert Laganière, University of Ottawa), can exhibit local movements that produce significant variations in the sensed image intensity. Yet, the image of such objects is to be assigned to the background rather than the foreground. Tree foliage produces intensity variations which resemble flicker due to the

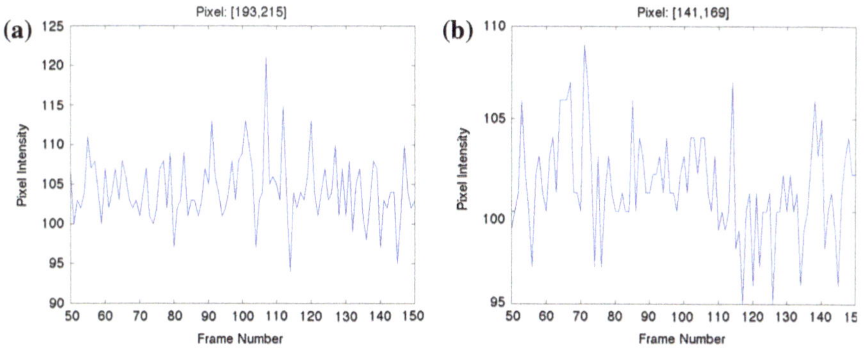

Fig. 4.2 Grey level temporal variations at two different pixels of the Building surveillance video sequence

Fig. 4.3 Staufer-Grimson motion detection method by mixture of Gaussians modelling of intensity: **a** The observed scene; everything is motionless in this scene except the tree foliage and the bike and its rider. **b** The detected foreground points. The points are on the image of the bicycle and its rider, and at the fringes of the tree foliage where the intensity variation between consecutive frames is more noticeable

continual fluctuations of the leaves which occur when air moves even slightly. These intensity variations can cause unwanted spurious motion to be detected (Fig. 4.3).

Image sequences acquired during rain, or snowfall (Fig. 4.1c) (image sequence from Karlsruhe University, Institut für Algorithmen und Kognitive Systeme, http://i21www.ira.uka.de/image_sequences/), are other examples where moving object detection is adversely affected. Conditions such as rain and snow produce generally incoherent temporal sensing measurements which can obscure the movement of targeted moving objects and thus jeopardize their detection.

Sensing noise always occurs and illumination change is prevalent in practice. Therefore, they must be included in any statement of motion detection. Occasional impediments which occur in some applications but not in others, such as unsought surface deformations of fixed environmental objects, as with tree foliage under blowing air, and atmospheric sensing interference, as with rain and snow, require specific processing which may include the intervention of procedures and descriptions external to the intrinsic process of motion detection.

Motion detection serves useful applications and offers technical and technological challenges worthy of research. It has been the subject of a considerable number of studies and a frequent theme in image processing conferences. It is also a long standing topic of computer vision [1, 2]. Key among applications are human activity analysis and traffic monitoring, principally for the impact they have or can have on collective and personal safety of people. There have been several lengthy surveys of investigations in these applications [3–9]. A general review of image change detection can be found in [10] and a survey of background modelling by Gaussian mixtures in [11]. Most of the research surveyed in these reviews looks at motion detection from the viewpoint of image change without an explicit link to the motion that caused it. However, although motion can be detected without referring to it explicitly, as in background differencing for instance, it remains the underlying fundamental variable. Moreover, there have been relatively very few studies which investigated variational formulations [12–18] when such statements are now common in computer vision, affording tractable accurate algorithms to solve difficult problems [19–22].

This chapter will focus on variational motion detection formulations. Variational methods, we recall, solve a problem by minimizing a functional which contains all of the problem variables and constraints. With the proper constraints, and the proper tools of optimization, these can lead to algorithms that are accurate in the evaluation of the problem unknowns, that are stable numerically, and whose behavior can be explained. The objective functionals of most of the variational formulations we will be discussing are active curve functionals minimized by developing the corresponding Euler-Lagrange equations and solving these using the level set formalism we have reviewed in Chap. 2.

The chapter sets out with a discussion of background modelling and background subtraction methods. Applications where reliable background models are workable, such schemes can be very efficient. The discussion is divided in two parts, one dealing with point-wise background subtraction (Sect. 4.2) and another treating variational, and global thereof, background subtraction (Sect. 4.3). Variational formulations explicitly reference the background and foreground as regions, thereby providing a natural description of the division of the image domain into a foreground and a background. Moreover, and contrary to point-wise methods, variational schemes can use spatial regularization to exploit the fundamental characteristic that neighbors in the image domain tend to have the same foreground/background membership, i.e., background points tend to cluster spatially, and so do the foreground points. Therefore, variational formulations can produce a detection map that is free of small, noisy regions that generally plague point-wise detection.

When viable background modelling is not practicable, detection can resort to consecutive image differencing or to the use of image motion variables such as optical flow or its normal component. These approaches are taken up in Sects. 4.4 and 4.5, respectively. The discussions up to Sect. 4.5 assume that the viewing system does not move so that the background is static modulo noise. Section 4.6 addresses the problem of detecting motion in the presence of viewing system movement. In this case, the image motion is the result of a superposition by vector addition of the motion due to the viewing system movement and to the moving environmental objects. The main strategy is to state the problem with the motion due to the viewing system subtracted from the overall image motion. This would bring the problem back to the case of moving objects against a static background so that the schemes of previous sections can be applied.

The discussions to follow assume that the image sequence is a spatiotemporal image intensity pattern, i.e., a temporal sequence of single valued images:

$$I(x, y, t) : \Omega \times]0, T[\rightarrow \mathbb{R}^+, \tag{4.1}$$

where Ω is the image domain and T is the duration of the sequence. However, the formulations can be simply generalized to multivalued images.

4.2 Background Modelling and Point-Wise Background Subtraction

The simplest of background models would be a snapshot of the environment when none of the anticipated motions occur, i.e., a template image of the environment at a time it contains only fixed constituent objects. Except in simple situations, one would easily expect this naive template to fail motion detection because it contains no information about the image temporal variations not caused by object motion, such as sensing noise or illumination change. This snapshot template model can be generalized by processing the temporal intensity stream at each point of the image domain, rather than a single template value, and describing the intensity variations in some window of time by a statistical model. Statistical models have been frequently used to advantage in variational image analysis, particularly image segmentation but also in motion analysis [22].

Statistical background intensity modelling can be examined from two distinct perspectives: *parametric* representation which uses a parametric distribution to describe the image intensity stream, and *nonparametric* representation which uses an empirical distribution (histogram) of the intensity.

4.2.1 Parametric Modelling: The Stauffer–Grimson Method

An immediate generalization of the naive template is the image averaged in a time interval [23]. This would be an improved template because averaging is a low pass filter which smooths out spurious variations of the sensed signal. This comes to assuming a Gaussian model with unit variance and a mean equal to the empirical average. For a slightly more general representation, the Gaussian model can include both the mean and the variance parameters. In practice, of course, the Gaussian density is discretized into bins of grey levels. The classification of a pixel has often been expedited in practice by comparing the value of the Gaussian density corresponding to its intensity to a threshold determined experimentally; the pixel is assigned to the background if this value is under the threshold, and to the foreground otherwise.

Although mathematically convenient, the Gaussian model is not generally applicable. For instance, it is not descriptive of non-additive noise and it is not expressive of scene illumination variations, specularity, and intensity changes due to changing local surface orientation of objects such as trees and water bodies. It is also not applicable when the background template must be constructed from an image sequence in the presence of moving objects because these objects might contribute arbitrary intensities to the background template at arbitrary position and time. All such conditions are common in applications, particularly outdoor scene surveillance and monitoring. To describe background templates under these circumstances, more descriptive models are needed. For instance, a mixture of Gaussians at each pixel has been used in [24]. The rationale is that sensing noise, illuminant brightness change, and transient background masking by moving objects, will cause distinct pixel intensity clusters, each represented by a distinct Gaussian distribution. The experimental investigations in [24], and elsewhere, support this assertion and show that K-means clustering is an expeditious substitute for Gaussian mixing.

The method in [24] may be implemented to run in real time and, as a result, it has been widely used in applications such as surveillance. It is not a variational method because it performs single pixel measurements, operations, and decisions, to classify each pixel $\mathbf{x}$ as part of the background or the foreground. More specifically, the scheme models the image intensity at a point $\mathbf{x}$ by a mixture of K Gaussians:

$$P(I(\mathbf{x})) = \sum_{i=1}^{K} c_i G(I(\mathbf{x})|\mu_i, \sigma_i), \tag{4.2}$$

where the means μ_i and the variances σ_i are estimated from the image values at $\mathbf{x}$, for each $\mathbf{x}$ of the image positional array, in a time interval ending with the current time. As we have already mentioned, the estimation can be expedited by K-Means clustering of the data at $\mathbf{x}$ into K groups each one of which is represented by a Gaussian. The mixture coefficients $c_i, i = 1, \ldots, K$ are approximated by the relative number data points in the clusters $i, i = 1, \ldots, K$, respectively.

Starting from an initial estimate, the mixture parameters are estimated iteratively as follows: The new image intensity value at $\mathbf{x}$ is mapped to the current clusters at $\mathbf{x}$. If it is an outlier, i.e., if it is too far, according to some threshold, from all K clusters, the farthest cluster representation according to the clusters Gaussian distributions is dropped and replaced by a Gaussian with the current image value as its mean, a high variance ("high" according to some reference value), and a low mixture coefficient value (according to some other reference value). Otherwise, the parameters of all distributions remain the same except for the distribution of the closest cluster j which is updated as follows: Its mixture coefficient is modified by:

$$c_j^n = c_j^{n-1} + \alpha \left(1 - c_j^{n-1}\right), \tag{4.3}$$

where n is the discrete time index, and α is a constant (called the learning rate). The representation mean is pulled toward the current image intensity value and its variance is modified accordingly using the formulas:

$$\begin{aligned} \mu_j^n &= \mu_j^{n-1} + \rho_j \left(I(\mathbf{x}) - \mu_j^{n-1}\right) \\ v_j^n &= v_j^{n-1} + \rho_j \left(\left|I(\mathbf{x}) - \mu_j^{n-1}\right|^2 - v_j^{n-1}\right), \end{aligned} \tag{4.4}$$

where ν_j indicates the variance, $\nu_j = \sigma_j^2$, and ρ_j is the learning factor, $\rho_j = \alpha G(I(\mathbf{x})|\mu_j, \sigma_j)$.

The update processing scheme of Eqs. (4.3) and (4.4) produces after some time a set of cluster representations of a mixture of Gaussians at each pixel, but has no provision for classifying observed image values, i.e., for assigning a pixel of the current image of the sequence to the foreground or background. In [24], the background model (which gives the foreground by complementarity), is determined by first ranking the clusters at $\mathbf{x}$ according to the values of c_i/σ_i and then the first m clusters for which $\sum_{i=1}^{m} c_i > L$, where L is a threshold, will be the background representation. The rationale for ordering the c_i/σ_i values is that the background is expected to have clusters with higher mixture coefficient values and lower variances.

Following the background/foreground representation by models everywhere on the image domain, the decision to classify a new image value at some point as background or foreground is done by mapping the value onto the clusters corresponding to the point. The method depends on heuristics, such as setting thresholds and other such constants, but these have natural explanations and can be dealt with by reasonable rules of thumb. The method has been used in many motion detection investigations and applications where it has generally performed well by succeeding in long-term learning of the intensity representation parameters and other necessary quantities. The scheme's output can be followed by connected component analysis [25] to extract explicitly the background/foreground regions.

Figure 4.3 (courtesy of Dr R. Laganière) shows an example which illustrates the kind of results that can be achieved with the scheme. The observed scene is depicted in (a). The bicycle and the tree foliage are moving and everything else is background. The camera is fixed. The detected foreground points are shown in (b). These are not organized into delimited regions, a process which is generally done with this scheme by connected component analysis. The foliage movement is detected mainly at its outer edges where the intensity from frame to frame varies significantly.

4.2.2 Nonparametric Modelling

Although the mixture of Gaussians offers a general representation of the image sequence at a point, it generally requires several components learned from a large data sample to be accurate. Learning can be quite time consuming. These are shortcomings which prompted [24] to simplify Gaussian mixture modelling to K-Means clustering. However, a more efficient alternative may be to use an empirical density, i.e., a histogram, to represent the image sequence at a pixel. Histograms have been effective representations in computer vision problems such as segmentation and tracking [26] as well as pattern recognition [27]. For the motion detection application, it can be learned simply by binning the sensed grey level intensities at a pixel within a time window ending at the current frame. A histogram is a direct record of the image sequence grey level profile. It can be continually updated to account for illumination change and transient masking by moving objects. At any current time, one can select the most often occurring value in the image histogram at $\mathbf{x}$ to be the background value at $\mathbf{x}$.

4.2.3 Image Spatial Regularization

Single pixel intensity modelling as discussed so far ignores the image spatial properties. There have been studies which addressed spatial consistency of pixel intensity models and its use in image interpretation. For instance, the Bayes/Markov random field regularization formalism was used in [28, 29] for moving object detection, and the graph cut data description and regularization framework was combined in [30] with clustering and Gaussian mixing for simultaneous segmentation and tracking of multiple objects.

In many applications an image can be approximated by a piecewise constant model, which means that it can be segmented into a set of regions in each one of which it is the sum of a constant plus random noise [19, 20, 22, 31–33], i.e., the model can be seen as an approximation of the image in which random perturbations such as sensing noise and artifacts have been smoothed out. As a result, this representation can be quite useful in motion detection since it removes or lessens intensity variations which are not due to object motion but which can be large enough to adversely affect the interpretation. The piecewise constant approximation of the image can be used as input to motion detection rather than the raw image. Alternatively, one can use the most often occurring image intensity at each point, taken from a piecewise constant approximation of the image rather than from the image directly; the map of the most often occurring grey level is expected to be more stable, i.e., to vary less in time than the raw image or its piecewise approximation and, therefore, be a better background model to use with background subtraction.

Variational formulations of piecewise constant image segmentation commonly involve minimizing a functional of two terms, a data term which measures the deviation of the image from the piecewise constant model approximation, and a regularization term for smooth segmentation region boundaries. The spatial smoothness of the approximating image data is explicit in the use of the constant image model representation, so that no additional term in the objective functional is necessary. The problem can be stated in the continuous domain [31] but it can also be solved directly and efficiently in the discrete case by graph cut optimization [34, 35], a scheme which we outline in the following.

Let $I : \mathbf{x} \in \Omega \subset \mathbb{R} \rightarrow I(\mathbf{x}) \in \mathscr{I}$ be a general image function from domain Ω to a space $\mathscr{I}$ of photometric variables such as intensity, intensity features, or colour. Graph cut methods state image segmentation as a label assignment problem. Partitioning of the image domain Ω amounts to assigning each pixel a label l in some finite set of labels $\mathscr{L}$. A region R_l is defined as the set of pixels with label l, i.e., $R_l = \{\mathbf{x} \in \Omega \mid \mathbf{x} \text{ is labeled } l\}$. The problem consists of finding the labelling which minimizes a given objective function describing some common constraints. Let λ be an indexing function which assigns each pixel to a region:

$$\lambda : \mathbf{x} \in \Omega \longrightarrow \lambda(\mathbf{x}) \in \mathscr{L}, \tag{4.5}$$

where $\mathscr{L}$ is the finite set of region labels. For instance, $\mathscr{L}$ can be the set of grey levels $\{0, \ldots, 255\}$ for a grey level image. The graph cut objective function, $\mathscr{F}$, can then be written as:

$$\mathscr{F}(\lambda) = \sum_{l \in \mathscr{L}} \sum_{\mathbf{x} \in R_l} \|\lambda(\mathbf{x}) - I(\mathbf{x})\|^2 + \alpha \sum_{\{\mathbf{x},\mathbf{y}\} \in \mathscr{N}} r_{\{\mathbf{x},\mathbf{y}\}}(\lambda(\mathbf{x}), \lambda(\mathbf{y})), \tag{4.6}$$

where $\|\cdot\|$ indicates the Euclidian norm, or the absolute value for scalar quantities, α is a positive constant to weigh the contribution of the first term (data term) relative to the other (regularization term), $\mathscr{N}$ is the pixel neighborhood set, and $r_{\{\mathbf{x},\mathbf{y}\}}(\lambda(\mathbf{x}), \lambda(\mathbf{x}))$ is a smoothness function, often in truncated form [34]:

$$r_{\{\mathbf{x},\mathbf{y}\}}(\lambda(\mathbf{x}), \lambda(\mathbf{y})) = \min(c^2, \|\mu_{\lambda(\mathbf{x})} - \mu_{\lambda(\mathbf{y})}\|^2), \tag{4.7}$$

where c is a constant threshold and μ_λ is a function characterizing the region labelled λ. For instance, when labels are grey levels then one can use $\mu_\lambda = \lambda$. The minimization of Eq. (4.6) can be done very efficiently by graph cut combinatorial optimization and implementations can be found on the web.

Figures 4.4a, b show an example of piecewise constant image representation by graph cut optimization. The original image is shown in (a) and the computed piecewise approximation in (b). Most grey levels in the range [0, 255] appear in the input image. Only about forty grey levels survive in the piecewise constant approximation, giving rise to forty constant grey level regions.

Figure 4.5 shows the grey scale temporal variations in the graph cut piecewise constant approximation of the building surveillance sequence, shown in Fig. 4.4, at the pixels of the raw image used in the graphs of Fig. 4.2. The variations have been reduced in magnitude because the pixels generally remain in the same region over

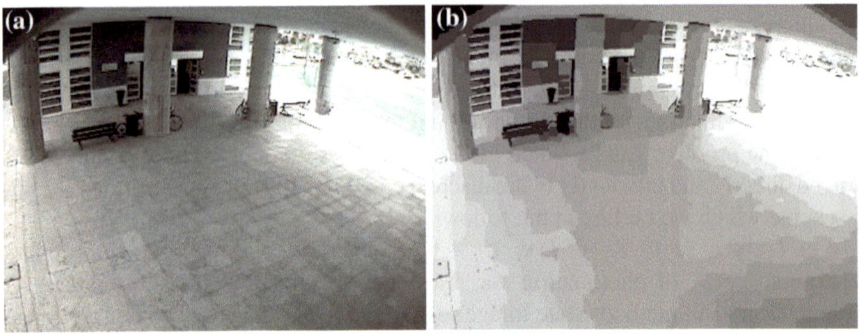

Fig. 4.4 Piecewise constant image approximation by graph cut representation and optimization: **a** The input image; **b** the piecewise constant approximation. The input image contains most of the grey levels in the range [0, 255] but the approximation has only about forty of these. The approximation will likely be more stable in time than the original image

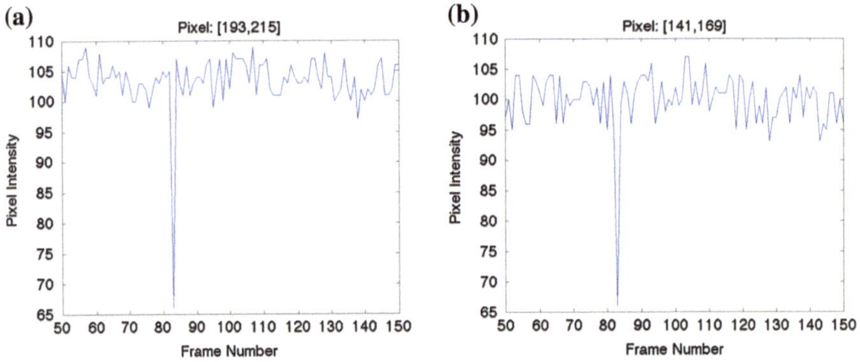

Fig. 4.5 Grey level temporal variations at pixels of the piecewise constant approximation by graph cut segmentation of the building surveillance video sequence shown in Fig. 4.4

time and the regularized grey level of the region varies moderately. It would be interesting to investigate whether this behavior is of a general nature.

4.3 Variational Background Subtraction

In the previous section we discussed background model building and corresponding point-wise background subtraction for motion detection. Single-pixel background/foreground detection does not exploit the spatial continuity of pixel classification, i.e., the fundamental property that neighbouring pixels tend to have the same interpretation. In contrast, variational formulations of background differencing refer explicitly to the background and the foreground as *regions* which partition the image domain.

In the following, we describe how a background model $\mathscr{M}$ can be used to detect motion by a variational method, i.e., a method where the desired partition corresponds to a minimum of a functional over a space of allowable image domain partitions. In the following, we will consider two distinct cases: when $\mathscr{M}$ is a probability model and when it is an image template.

4.3.1 Probability Models

Let $\mathscr{M}(\mathbf{x})$ be a probability model of the background at $\mathbf{x} \in \Omega$. Motion detection by background subtraction can be stated as a two-region image partitioning problem by maximizing the following functional, $\mathscr{E}$, containing a data term which constrains the region representing the background to conform to probability model $\mathscr{M}$, and a term for the region representing the foreground and related to a threshold on the conformity of the image to the model. This is a region-based functional because it refers to region information, in contrast to region boundary information.

$$\mathscr{E}(R) = \int_{R^c} P(I(\mathbf{x})|\mathscr{M}(\mathbf{x}))d\mathbf{x} + \lambda \int_{R} d\mathbf{x}, \tag{4.8}$$

where R represents the foreground and its complement R^c the background. The coefficient λ in such a linear combination of terms is generally interpreted simply as a weight modulating the contribution of the term it multiplies. In the case of this functional, coefficient λ has a more explicit interpretation as a threshold which the image intensity probability must exceed at background points. We will show this after we derive the Euler-Lagrange equations for the maximization of Eq. (4.8), which we will do by rewriting the functional via the active curve formalism as follows.

Let $\gamma(s) : s \in [0, 1] \to \mathbf{x}(s) \in \Omega$ be a closed simple parametric curve of the plane and R_γ its interior. The objective functional can be rewritten as:

$$\mathscr{E}(\gamma) = \int_{R_\gamma^c} P(I(\mathbf{x})|\mathscr{M}(\mathbf{x}))d\mathbf{x} + \lambda \int_{R_\gamma} d\mathbf{x}. \tag{4.9}$$

By embedding γ in a one-parameter family of curves indexed by algorithmic time τ, $\gamma(s, \tau) : s, \tau \in [0, l] \times \mathbb{R}^+ \to (\mathbf{x}(s, \tau), \tau) \in \Omega \times \mathbb{R}^+$, we can determine the functional derivative of the objective functional Eq. (4.9) with respect to γ:

$$\frac{\partial \mathscr{E}}{\partial \gamma} = (\lambda - P(I|\mathscr{M}))\, \mathbf{n}, \tag{4.10}$$

where $\mathbf{n}$ is the outward unit normal function of γ. This gives the Euler-Lagrange ascent equation to drive γ to a maximum of Eq. (4.9):

$$\frac{\partial \gamma}{\partial \tau} = (\lambda - P(I|\mathscr{M}))\,\mathbf{n} \tag{4.11}$$

Curve γ is called an active contour under evolution equation (4.11). The motion of γ is along $\mathbf{n}$ at each of its points, and its speed function is $(\lambda - P(I|\mathscr{M}))$. The desired partition into background and foreground is given by the curve at convergence, i.e., as $\tau \to \infty$.

By examining the speed function at a point $\mathbf{x}$, $V = (\lambda - P(I(\mathbf{x})|\mathscr{M}))$, we note that the curve will move to assign $\mathbf{x}$ to region R_γ^c representing the background if $P(I(\mathbf{x})|\mathscr{M}) > \lambda$ and to region R_γ representing the foreground otherwise. Therefore, λ can be seen as a classification threshold: when $P(I(\mathbf{x})|\mathscr{M}) < \lambda$, it is better to assign $\mathbf{x}$ to the foreground. Otherwise it is classified as part of the background. However, this threshold is not applied pixel-wise and independently for different pixels but in the context of optimizing the global objective functional Eq. (4.9).

One can add a curve length term $\mathscr{S}$ to the objective functional so as to promote smooth region boundaries [31, 36, 37]:

$$\mathscr{S}(\gamma) = -\beta \int_\gamma ds, \tag{4.12}$$

the functional derivative of which is:

$$\frac{\partial \mathscr{S}}{\partial \gamma} = -\beta\kappa\mathbf{n}. \tag{4.13}$$

Coefficient β weighs the contribution of the boundary smoothness term against the data fidelity term. With the length term, the curve evolution equation becomes:

$$\frac{\partial \gamma}{\partial \tau} = (\lambda - P(I|\mathscr{M}) - \beta\kappa)\,\mathbf{n} \tag{4.14}$$

Equation (4.14) can be implemented by an explicit discretization of γ using a number of marker points. However, and as we have discussed in the preliminaries of Chap. 2, such an implementation faces serious problems. First, topological changes, fans, and shocks, which can occur during curve evolution, cannot be processed in general. Second, results depend on the parametrization and errors in the representation can significantly grow cumulatively during evolution. A better method is to represent γ implicitly as the zero level-set of a function ϕ, called a level set function, $\phi : \mathbb{R}^2 \to \mathbb{R}$, i.e., γ is the set $\{\phi = 0\}$. By evolving the level set function ϕ, rather than the curve, the topological variations of the curve occur automatically and its position at any time can be recovered as the level zero of ϕ. The level set method has been reviewed in Chap. 2 and there is an extensive literature on effective numerical algorithms to implement level set evolution equations [38].

As reviewed in Chap. 2, when a curve moves according to $\frac{d\gamma}{d\tau} = V\mathbf{n}$, the level set function evolves according to:

$$\frac{\partial \phi}{\partial \tau}(\tau) = V \|\nabla \phi\|. \tag{4.15}$$

Assuming ϕ positive in the interior of γ and negative in the exterior, the outward unit normal $\mathbf{n}$ and the curvature κ in Eq. (4.14) can be expressed in terms of the level set function ϕ:

$$\mathbf{n} = -\frac{\nabla \phi}{\|\nabla \phi\|} \tag{4.16}$$

and

$$\kappa = \operatorname{div}\left(\frac{\nabla \phi}{\|\nabla \phi\|}\right) \tag{4.17}$$

In our case, the velocity V is given by the right-hand side of Eq. (4.14). Therefore, the corresponding level set function evolution equation is given by:

$$\frac{\partial \phi}{\partial \tau} = (\lambda - P(I|\mathcal{M}) - \beta\kappa)\,\|\nabla \phi\| \tag{4.18}$$

Although Eq. (4.14) refers to the points on γ, the velocity can be computed everywhere in Ω. Therefore, Eq. (4.18) can be used to update the level set function everywhere in Ω.

As discussed in Sect. 4.2, probability models can be approximated from data by assuming a parametric form of the density and estimating the parameters, or by nonparametric estimates such as histograms.

4.3.2 Template Models

A background template model is typically chosen to be a view of the scene containing only objects that are static during the detection process, i.e., fixed objects that are usually part of the scene. Given a current image, one can then look at motion detection in one of two distinct fundamental ways, as determining *regions* where the difference between the image and the template at corresponding points is high, or as determining the *boundaries* where the gradient of this difference is strong. A combination of both ways is of course possible and can be of benefit to some applications.

4.3.2.1 Region-Based Template Subtraction Detection

Region-based motion detection using a background template can be formulated as the minimization of the following functional [12]:

$$\mathcal{E}(\gamma) = \int_{R_\gamma^c} ((B - I)^2 d\mathbf{x} + \lambda \int_{R_\gamma} d\mathbf{x}, \tag{4.19}$$

where B is the background template image. This functional uses directly the difference between the background and the image rather than the probability of the image according to the background probability model [as such, we minimize the functional rather than maximize it as with Eq. (4.9)]. However, this amounts to assuming that the difference between the background template and the image to be zero-mean, unit-variance Gaussian. Indeed, the data function is, up to an additive positive constant, equal to the negative of the logarithm of the zero-mean, unit-variance normal density function, therefore a special case of general model-based data functions discussed in the preceding function.

Coefficient λ in Eq. (4.19) may, as in Eq. (4.9), be interpreted as a threshold, but this time on the data function $(B - I)^2$ to decide whether there has been a change or not in the image compared to its template: if $(B - I)^2 > \lambda$ at $\mathbf{x}$, it is better to assign the point to the foreground.

If we add a length regularization term to the objective functional, to bias detection toward a partition with a smooth foreground/background boundary:

$$\mathscr{S} = \beta \int_{\gamma} ds, \tag{4.20}$$

where β is a positive constant, the minimization, conducted as before, leads to the following Euler-Lagrange descent curve evolution equation:

$$\frac{\partial \gamma}{\partial \tau} = -\left(\lambda - (B - I)^2 + \beta\kappa\right)\mathbf{n}. \tag{4.21}$$

The corresponding level set evolution equation is given by:

$$\frac{\partial \phi}{\partial \tau} = -\left(\lambda - (B - I)^2 + \beta\kappa\right)\|\nabla\phi\|. \tag{4.22}$$

Example : This is an example from a camera surveillance application (courtesy of Dr Hicham Sekkati). The monitored scene is a bank ATM booth. The template is shown in Fig. 4.6a. It is an image of the ATM booth interior environment with only the static objects normally appearing in it. Figure 4.6b shows an image of the current scene, with a person using the machine. A grey level depiction of the difference between the template and the current image is displayed in Fig. 4.7a and the detected motion region, which is the interior of the active curve at convergence, is depicted in Fig. 4.7b. The detected region has positioned the person properly even though its boundary does not bear much precise shape information. However, in applications where the moving objects need to be properly located but without necessarily having an accurate delineation of their boundary such results can be very useful.

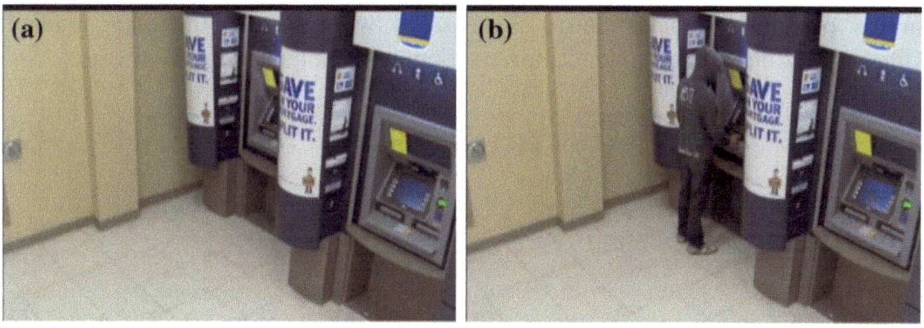

Fig. 4.6 Motion detection by background subtraction: **a** the template image of the scene; it contains only the objects that are normally part of the scene and are static during the detection process and, **b** an image of the current scene, in which motion detection is to be performed

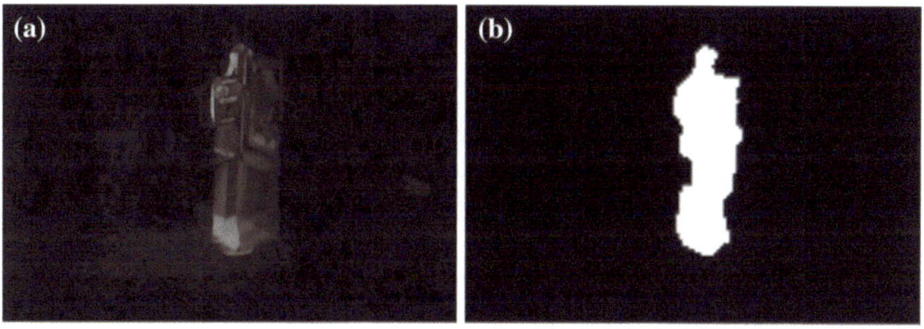

Fig. 4.7 Motion detection by region-based background subtraction via level set evolution Eq. (4.22): **a** grey-level representation of the background subtraction image and, **b** the region detected as the foreground. The detected foreground contains useful information about the position and shape of the detected object (person), even though its boundary is only a sketch of the object occluding contour

4.3.2.2 Boundary-Based Template Subtraction Detection

The *geodesic active contour* functional [39, 40] is a boundary-based integral functional originally presented as a means to detect in images object contours of strong intensity contrast. Its general form is, for a parametric curve γ defined on $[a, b]$:

$$\mathscr{F}(\gamma) = \int_a^b g\left(\|\nabla I(\gamma(q))\|\right) \|\gamma'(q)\| dq, \tag{4.23}$$

where ∇I is the image spatial gradient, and $g : [0, \infty] \to \mathbb{R}^+$ is a positive monotonically decreasing function verifying the condition:

$$\lim_{z \to +\infty} g(z) = 0. \tag{4.24}$$

In this formulation, image contrast is measured by the norm of the image gradient. Other definitions are, of course, allowed. Common choices of function g are:

$$g(z) = \frac{1}{1+z^2} \tag{4.25}$$

and

$$g(z) = e^{-z^2} \tag{4.26}$$

with such a function g, one can easily see that the quantity $g(\|\nabla I\|)$ is an edge indicator in the sense that it is low at strong edges and high at weak ones. We will see shortly that the geodesic active contour velocity is mediated by $g(\|\nabla I\|)$ and its gradient, in such a way that it will slow down at object boundaries and adhere to them.

The geodesic functional is parametrization independent because if f is a re-parametrization function:

$$f : r \in [c, d] \rightarrow q = f(r) \in [a, b]; \quad f' > 0, \tag{4.27}$$

we have:

$$\mathscr{F}(\gamma \circ f) = \int_c^d g\left(\|\nabla I(\gamma \circ f)(r)\|\right) \|(\gamma \circ f)'(r)\| dr, \tag{4.28}$$

where $\circ$ indicates composition of functions. Therefore, the Euler-Lagrange equations corresponding to the minimization of the functional would remain the same. When the curve is parametrized by arc length, the functional is written as:

$$\mathscr{F}(\gamma) = \int_0^l g\left(\|\nabla I(\gamma(s))\|\right) ds, \tag{4.29}$$

where s designates arc length and l the Euclidean length of γ; in shorthand notation, it can be written:

$$\mathscr{F}(\gamma) = \int_\gamma g\left(\|\nabla I\|\right) ds \tag{4.30}$$

The curve evolution equation to minimize a geodesic functional $\mathscr{F}$ can be derived in the usual way, first by embedding γ in a one-parameter family of curves $\gamma : s, t \in [0, l] \times \mathbb{R}^+ \rightarrow \gamma(s, \tau) \in \Omega \times \mathbb{R}^+$, and adopting the Euler-Lagrange descent equation

$$\frac{\partial \gamma}{\partial \tau} = -\frac{\partial \mathscr{F}}{\partial \gamma}, \tag{4.31}$$

The functional derivative with respect to γ of the integral over γ of a positive scalar function $h = h(\mathbf{x}(s))$ is given by (Chap. 2, [41]).

$$\frac{\partial \int_\gamma h ds}{\partial \gamma} = (< \nabla h, \mathbf{n} > -h\kappa)\mathbf{n}, \tag{4.32}$$

where $< .,. >$ denotes the scalar product. Therefore, this gives the following movement at every point on the geodesic curve γ:

$$\frac{\partial \gamma}{\partial \tau} = \left(g\left(\|\nabla I\|\right)\kappa - < \nabla g\left(\|\nabla I\|\right), \mathbf{n} >\right)\mathbf{n} \tag{4.33}$$

This evolution equation will drive the curve to adhere to high contrast object boundaries, i.e, object contours of high intensity transitions. The curvature term of the velocity vector, with a corresponding speed equal to curvature modulated by the positive function g, promotes shorter, smoother curves. At ideal edges, i.e., where $\|\nabla I\| \to \infty$, we have $g \to 0$. For this reason, g is often called a *stopping function*. In practice, g is small at edges and the contribution of the first term to the curve speed weakens significantly. The second velocity vector component, sometimes called a *refinement term*, drives the curve towards significant edges because the gradient vector ∇g points toward high image transitions (Fig. 4.8). This component assists the first term to effectively inhibit curve evolution at high contrast object boundaries. Its contribution is essential when the stopping function is not sufficiently low everywhere on the target object boundaries.

Both velocity terms depend on the image gradient, which will cause the curve to linger at open, irrelevant high contrast images boundaries. Therefore, a geodesic evolution can be slow reaching the desired object boundaries. It is customary in this case to add a *balloon velocity term* [39] to speed it up:

$$-\nu g\left(\|\nabla I(\boldsymbol{\gamma})\|\right)\mathbf{n} \tag{4.34}$$

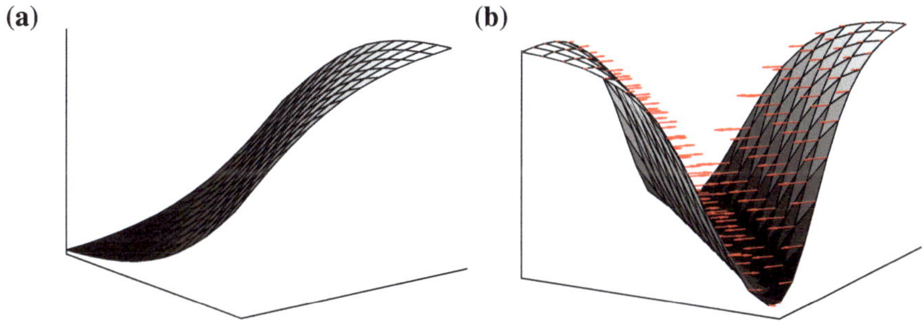

Fig. 4.8 Geometric interpretation in two dimensions of the geodesic active curve velocity vector profile at a contrast edge: **a** a 2D ramp edge, **b** the corresponding valley created by the stopping function g. The speed of the curve, in the direction of the curve normal, is $g\left(\|\nabla I\|\right)\kappa - < \nabla g\left(\|\nabla I\|\right), \mathbf{n} >$. At a strong edge, the term $g\left(\|\nabla I\|\right)\kappa$ is very small because $g \to 0$ when $\|\nabla I\| \to \infty$. The refinement speed $- < \nabla g\left(\|\nabla I\|\right), \mathbf{n} >$ attracts the evolving contour to the valley because the gradient vector ∇g points toward high image transitions

This is the velocity one would get from the functional derivative of the region term:

$$\nu \int_{R_\gamma} g\left(\|\nabla I\|\right) d\mathbf{x} \tag{4.35}$$

Coefficient ν is chosen positive when the curve encloses the target object and is made to move inward. It is negative when, instead, the curve is initially enclosed by the desired object and is made move outward.

The level set evolution equation corresponding to Eq. (4.33) is:

$$\frac{\partial \phi}{\partial \tau} = \kappa g\left(\|\nabla I\|\right) \|\nabla \phi\| - < \nabla g\left(\|\nabla I\|\right), \nabla \phi > \tag{4.36}$$

If a balloon component is included then:

$$\frac{\partial \phi}{\partial \tau} = \kappa g\left(\|\nabla I\|\right) \|\nabla \phi\| - < \nabla g\left(\|\nabla I\|\right), \nabla \phi > - \nu g(\|\nabla I\|)\|\nabla \phi\| \tag{4.37}$$

An important property of the geodesic contour functional $\mathscr{F}$ is that it is, as we have seen, curve parametrization invariant, i.e., a re-parametrization of γ does not affect it, contrary to its precursor, the Snake active curve functional of [42] which is parametrization dependent. With a parametrization dependent functional, a change in the curve parametrization can lead to a different curve evolution and, therefore to different results. Another significant advantage of the geodesic functional over the Snake is that it is amenable to the level set implementation, contrary to the Snake which relies on an explicit curve representation as a set of points which are explicitly moved, exposing the evolution to irreparable numerical ills. The level set representation, we know (Chap. 2), secures a stable curve evolution execution.

For our motion detection application we will, of course, use a geodesic driven not by the strength of the image gradient but, rather, by the strength of the difference between the current image and the template, measured, for instance, by $\|\nabla(B-I)\|$, in which case the geodesic would be:

$$\mathscr{F}(\gamma) = \int_\gamma g\left(\xi\right) ds, \tag{4.38}$$

where $\xi = \|\nabla(B - I)\|$. One may also use $\xi = |B - I|$.

Example : This is the bank ATM booth scene shown in Fig. 4.6. A grey level depiction of the absolute difference between the template and the current image is reproduced again in Fig. 4.9a and the detected motion region, the interior of the geodesic in its position at convergence, is depicted in Fig. 4.9b. As with the preceding region-based method of detection by background subtraction, this geodesic detected foreground contains useful information about the position and shape of the moving object, even though its boundary is only a sketch of the object occluding contour.

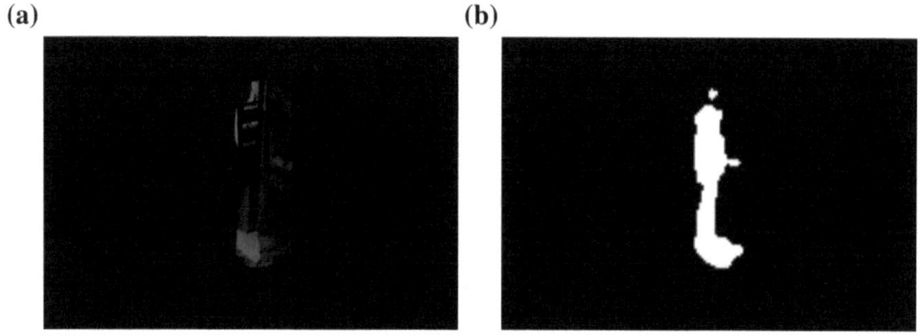

Fig. 4.9 Motion detection by a background subtraction geodesic: **a** grey-level representation of the background subtraction image and, **b** the detection region corresponding to the motion boundary detected by the geodesic. The detected foreground contains useful information about the position and shape of the moving object, even though its boundary is only a sketch of the object occluding contour

4.4 Detection by Image Differencing

Detection by image differencing uses the amplitude of the intensity difference between two consecutive images, or a combination of such, to determine the region of the image domain which corresponds to objects moving in the observed environment. The problem can be looked at as detection by background template differencing where the first of the two consecutive images plays the role of the template. However, there are important differences.

First, there is the fact that consecutive video images generally occur in a very short interval of time, implying that the region covered in one image by slow moving objects may overlap significantly the region the objects occupy in the other image. If these object regions have no significant image spatial variations, i.e., are not textured, they may not be detected, in the sense that only the small image areas covered/uncovered by motion will be accessible to detection. This situation does not generally occur with the background differencing scheme. If, instead, the motion between frames is significant, so that the areas uncovered and covered by object motion do not overlap, image differencing will produce at each frame two regions for the moving object, one corresponding to the covered image and the other to the uncovered, which is not the correct foreground since it should be only one of these two regions. This also does not occur with background differencing detection.

The second difference, just as important as the first, is that for small image motion occurring in a short period of time, the difference of two consecutive images is an approximation of the temporal image derivative $I_t = \frac{\partial I}{\partial t}$, a quantity which can be analytically related to motion, in contrast with the difference between an image and its background template. This relation is conveyed by the Horn and Schunck optical flow equation [43]:

$$I_x u + I_y v + I_t = 0, \tag{4.39}$$

where $(I_x, I_y) = \nabla I$ is the image spatial gradient, and $W = (u, v)$ is the image motion, or optical flow. From Eq. (4.39) we get:

$$I_t = -\|\nabla I\| W^{\perp}, \tag{4.40}$$

where $W^{\perp} = W \cdot \frac{\nabla I}{\|\nabla I\|}$ is the component of W is the direction of the gradient, i.e., in the direction perpendicular to the isophote. We know from Chap. 2 that this is a manifestation of the aperture problem because the image data at individual points gives access not to the full motion but only to its normal component which, as a result, can be appropriately called the *visible motion*. According to Eq. (4.40), the temporal derivative is the visible motion modulated by image contrast. Therefore, it is a good motion cue, and detection by image differencing which uses it is a sensible scheme. However, it is important to remember that this interpretation of I_t is valid for small intervals of time and small motions. It is also important to remember that not all image temporal variations are due to motion as we have discussed in Sect. 4.1.

A straightforward variational formulation to detect motion by image differencing consists of minimizing a region-based functional similar to the background template differencing Eq. (4.19). Because the moving object regions in one image and the other can overlap significantly, the overlap can be missed by detection when the objects do not have sufficient texture. In this case, it may be advantageous to use a *boundary-based* functional which would look for the moving objects boundary where, in general, there is sufficient image differencing contrast. We will examine both approaches. In the subsequent discussions on image differencing, we will use the continuous notation I_t to designate the difference $(I_2 - I_1)$ between two consecutive images.

4.4.1 Region-Based Image Differencing Detection

If one looks at the foreground of moving objects in an image as the region where the squared difference between consecutive images exceeds a given threshold, motion detection can be done by minimizing the following objective functional:

$$\mathscr{E}(\gamma) = \int_{R_\gamma^c} I_t^2 \, d\mathbf{x} + \lambda \int_{R_\gamma} d\mathbf{x}, \tag{4.41}$$

where, as before, γ is a simple closed plane curve, with the interior R_γ representing the foreground and R_γ^c the background. Coefficient λ can be interpreted as a threshold which I_t^2 must exceed in the motion foreground, because the functional can be minimized by the following classification rule:

$$\begin{cases} \mathbf{x} \in \text{background for } I_t^2 < \lambda \\ \mathbf{x} \in \text{foreground} \quad \text{otherwise} \end{cases}$$

Augmenting Eq. (4.41) with a curve length term will add regularization to the classification from this decision rule with the practical effect of removing small, isolated regions from the foreground/background partition, such as those noisy sensing can produce:

$$\mathscr{E}(\gamma) = \int_{R_\gamma^c} I_t^2 \, d\mathbf{x} + \lambda \int_{R_\gamma} d\mathbf{x} + \alpha \int_\gamma ds. \tag{4.42}$$

The curve evolution equation to minimize Eq. (4.42) is:

$$\frac{\partial \gamma}{\partial \tau} = -\left(\lambda - I_t^2 + \alpha\kappa\right)\mathbf{n}, \tag{4.43}$$

and the corresponding level set evolution equation is, with the orientation and sign conventions described in the section on level sets of Chap. 2:

$$\frac{\partial \phi}{\partial \tau} = -\left(\lambda - I_t^2 + \alpha\kappa\right)\|\nabla\phi\| \tag{4.44}$$

Example : With an unmoving viewing system, consecutive image differencing is a sensible means of motion detection, but one needs to remember that the detected foreground will contain the areas covered by the moving objects in both the first and the second of the two successive images used. This can be an important consideration when the object movements are of large extent. The following example illustrates this fact. It uses two consecutive images of a person entering an office (courtesy of Dr Hicham Sekkati). The camera is static. The second image is shown in Fig. 4.10a. The movement is of large magnitude so that the images of the person in the two images have little overlap. Expectedly so, the detected foreground includes both regions covered by the person in the first and second image. It also contains a small segment of the table where there actually is some important intensity change between the images due to shadowing in the first image but not in the second.

(a) **(b)**

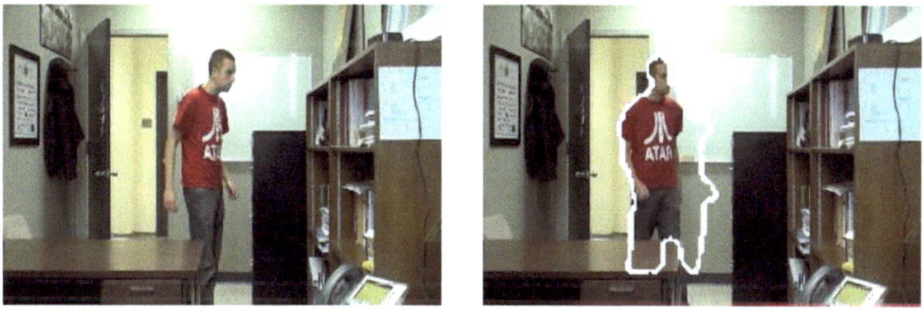

Fig. 4.10 Motion detection by region-based image differencing via the minimization of Eq. (4.41): **a** the second of the two consecutive images used and, **b** the boundary of the detected region superimposed on the first image. The detected motion region includes, expectedly so, both regions covered by the object in the first and second image

4.4.2 MAP Image Differencing Detection

The problem of region-based motion detection by image differencing can be cast in a Bayesian formulation [13]. Let $\mathscr{R} = \{R, R^c\}$ be a partition of the image domain Ω into foreground (R) and background (R^c). The problem can be stated as a maximum a posteriori estimation (MAP) of R: Among all allowable partitions, determine the most probable given the temporal image variation I_t, i.e., determine R such that:

$$\tilde{R} = \arg\max_R P(\mathscr{R}|I_t) = \arg\max_R P(I_t|\mathscr{R})P(\mathscr{R}) \tag{4.45}$$

Assuming conditional independence of the image difference measurements at points $\mathbf{x} \neq \mathbf{y}$, we have:

$$P(I_t|\mathscr{R}) = \prod_{\mathbf{x}\in R} P\left(I_t(\mathbf{x})\,|\mathscr{R}\right) \prod_{\mathbf{x}\in R^c} P\left(I_t(\mathbf{x})|\mathscr{R}\right). \tag{4.46}$$

Therefore, the MAP estimation of the foreground R can be stated as:

$$\tilde{R} = \arg\min_R \mathscr{E}(R) \tag{4.47}$$

with

$$\begin{aligned} \mathscr{E}(R) = &- \int_{\mathbf{x}\in R} \log P(I_t(\mathbf{x})|\mathscr{R})d\mathbf{x} \\ &- \int_{\mathbf{x}\in R^c} \log P(I_t(\mathbf{x})|\mathscr{R})d\mathbf{x} - \log P(\mathscr{R}). \end{aligned} \tag{4.48}$$

The first two integral terms on the right-hand side of Eq. (4.48) measure the agreement between the partition and the image difference measurements, according to a probability distribution to be specified. The third term is the prior on the partition, also to be specified. In [13], the following types of probability distribution models were used:

$$P(I_t(\mathbf{x})|\mathscr{R}) \propto \begin{cases} e^{-\frac{\alpha}{1+|I_t|}} & \text{for } \mathbf{x} \in R \\ e^{-\beta|I_t|} & \text{for } \mathbf{x} \in R^c, \end{cases}$$

where $\propto$ is the "proportional to" symbol and α, β are positive constants. Essentially, these models will bias motion detection toward partitions in which the foreground has high absolute image differences and the background has low differences. The prior can be simply modelled by, for some positive μ:

$$P(\mathscr{R}) \propto e^{-\mu \int_{\partial R} ds}, \tag{4.49}$$

where ∂R is the boundary of R, assumed regular. This model will favor shorter, therefore smoother, foreground boundaries, thereby promote the removal of small noisy fragments in the partition. With these models, the problem becomes: minimize

$$\mathscr{E}(R) = \alpha \int_R \frac{1}{1 + |I_t|} d\mathbf{x} + \beta \int_{R^c} |I_t| \, d\mathbf{x} + \mu \int_{\partial R} ds, \tag{4.50}$$

where α, β, and μ are constants to weigh the contribution of the terms they multiply.

Let $\gamma(s) : [0, 1] \rightarrow \Omega$ be a closed simple parametric plane curve to represent ∂R, where s is arc length and $R = R_\gamma$ is the interior of γ. By embedding γ in a one-parameter family of curves $\gamma(s, \tau) : [0, 1] \times \mathbb{R}^+ \rightarrow \Omega \times \mathbb{R}^+$, we have the following curve evolution equation to minimize Eq. (4.50) and, therefore, determine a partition of the image domain into foreground and background:

$$\frac{\partial \gamma}{\partial \tau} = -\left(\alpha \frac{1}{1 + |I_t|} - \beta |I_t| + \mu \kappa \right) \mathbf{n}, \tag{4.51}$$

where, as before, $\mathbf{n}$ is the outward unit normal function of γ and κ is its curvature function. The corresponding level set evolution equation is

$$\frac{\partial \phi}{\partial \tau} = -\left(\alpha \frac{1}{1 + |I_t|} - \beta |I_t| + \mu \kappa \right) \|\nabla \phi\|. \tag{4.52}$$

The coefficients α, β, μ, which can be normalized to add to 1, must be set appropriately for proper algorithm behavior. Coefficient μ affects the smoothness of the evolving curve. A version of this formulation which includes the case of a moving viewing system will be described in Sect. 4.6.

Example : The MAP image differencing model Eq. (4.50) has been investigated in the spatiotemporal domain to implement a motion tracking scheme [13]. The result shown in Fig. 4.11 is from an example in [13] using a slightly different input, namely the difference image scaled by the image gradient magnitude rather than the difference image, and a term characteristic of motion boundaries. The scene, recorded in a 24 f/s low resolution sequence, shows a person picking up a bag on the floor by stooping while walking. The bag becomes part of the foreground when picked. An extended spatiotemporal formulation which accounts for camera motion will be described in Chap. 5.

4.4.3 Boundary-Based Image Differencing Detection

With boundary-based image differencing detection, we detect motion boundaries via consecutive image difference boundaries rather than via image contrast boundaries. If a moving object boundary is looked at as an image contour of high consecutive image difference then minimizing the following objective functional is relevant to motion detection:

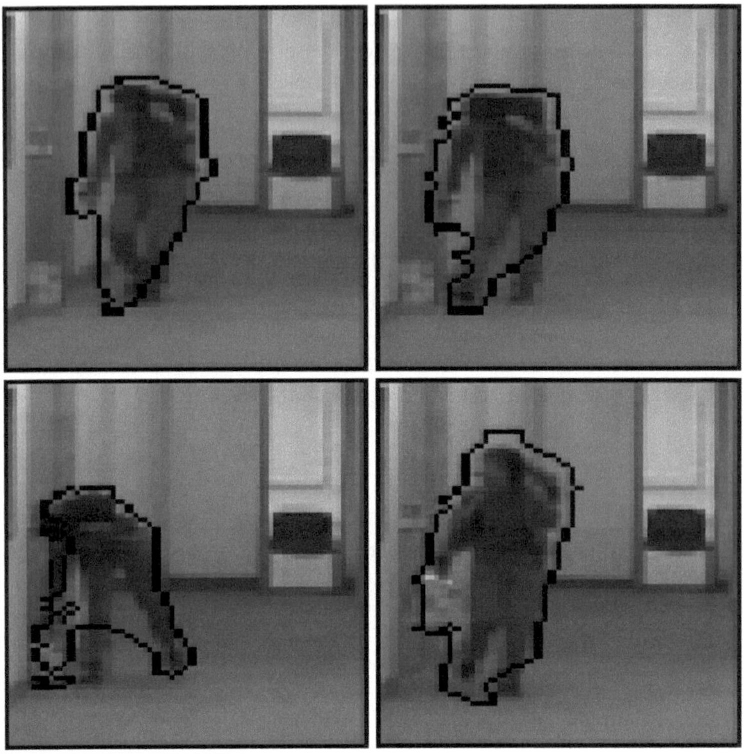

Fig. 4.11 *Picking bag* sequence (from [13]): An initial curve is placed wide so as to contain the foreground, and made to move to close on the foreground, first by containing moving parts of the person reaching for the back and then to also include the bag once picked

$$\mathscr{F} = \int_{\gamma} g\left(|I_t|\right) ds, \tag{4.53}$$

where g is defined as before. A moving object boundary may also be looked at as a contour of high image difference gradient, in which case the following functional can be used:

$$\mathscr{F} = \int_{\gamma} g\left(\|\nabla I_t\|\right) ds \tag{4.54}$$

Moving object contours can also be characterized in others ways. For instance, the investigations in [13, 44] used a function of second order image derivatives which is theoretically zero everywhere for piecewise constant image motion fields except at motion boundaries. This function is obtained as follows. Assuming that everywhere except at motion boundaries the motion field is locally approximately constant, i.e., $\nabla u \approx 0$ and $\nabla v \approx 0$, and that the image I is twice continuously differentiable, differentiation with respect to the image spatial coordinates of the Horn and Schunck optical flow constraint:

$$< \nabla I, W > + I_t = 0, \tag{4.55}$$

yields

$$\nabla (< \nabla I, W > + I_t) \approx \mathbf{H}W + \nabla I_t \ \approx \mathbf{0}, \tag{4.56}$$

where $\mathbf{H}$ is the Hessian matrix of image I, i.e., the matrix of second order derivatives of I.

$$\mathbf{H} = \begin{pmatrix} \frac{\partial^2 I}{\partial x^2} & \frac{\partial^2 I}{\partial x \partial y} \\ \frac{\partial^2 I}{\partial x \partial y} & \frac{\partial^2 I}{\partial y^2} \end{pmatrix} \tag{4.57}$$

Combining Eq. (4.56) with Eq. (4.55), we obtain the following motion boundary characterization function:

$$h = \left| \det(\mathbf{H}) I_t - < \nabla I, \left(\mathbf{H}^* \nabla I_t \right) > \right|, \tag{4.58}$$

where $|\cdot|$ denotes the absolute value; $\mathbf{H}^*$ is the transpose of the matrix of cofactors of $\mathbf{H}$ (it has the property $\mathbf{H}^*\mathbf{H} = \det(\mathbf{H})\mathbf{I}$, where $\mathbf{I}$ is the identity matrix). Function h is an indicator of motion boundaries because it takes small values inside motion regions where motion is assumed smooth, and generally large values at motion boundaries where the optical flow constraint equation is theoretically not valid.

Using function h, motion detection can be done by minimizing a geodesic functional of the form:

$$\mathscr{F}(\gamma) = \int_\gamma g(h) ds, \tag{4.59}$$

where g is, for instance, given by Eq. (4.25).

In motion detection applications, we do not know, by definition, where the moving objects are. Therefore, and because a geodesic active contour moves in a single direction [39], either inward or outward, the initial curve must be positioned wide enough so as to surround all the moving objects. In practice, one would place the initial curve close to the image domain boundary.

A ballon term can be added to the functional to speed up detection. The balloon velocity would be, where coefficient ν is positive for an inward moving geodesic:

$$-\nu g \left(|I_t| \right) \mathbf{n}, \tag{4.60}$$

Because g is a monotonically decreasing function, this balloon velocity is larger at points with lower image differences, i.e., the ballon term makes the curve move faster where there likely is no moving object boundary.

Geodesic curves can leak through a moving object boundary at places, called holes, where the argument of the boundary function is small. In the motion detection application of image differencing, these are segments of the moving object boundaries where the intensity difference between consecutive images is faint, a condition that would be present when the intensity transition between the background and the moving object is flat or blurred. Leakage can be abated by adding a properly weighed length regularization term to the geodesic functional. The purpose of using this term is to increase the value of the objective functional when the curve extends past a target boundary by encroachment through a hole. However, setting the weight value of the length regularization term can be difficult because such a term has the effect of shrinking the active curve overall, i.e., the curve can be moved through the target boundary as a result of a weight value improperly set too high. Leakage can be abated more securely by the addition to the objective functional of a region term which characterizes the image in the interior of the target object [45]. This characterization can be photometric, i.e., related to the moving objects intensity profile, or motion-based, i.e., associated with properties of the object motion. In the next section, we will see an example of a motion-based region term used in conjunction with the image contrast geodesic term.

In general, it is beneficial, for accrued detection robustness, to have an objective functional which contains both region-based and geodesic terms. For instance,

$$\mathscr{E} = \int_{R_\gamma^c} I_t^2 \, d\mathbf{x} + \lambda \int_{R_\gamma} d\mathbf{x} + \alpha \int_\gamma ds + \int_\gamma g\left(|I_t|\right) ds \tag{4.61}$$

Example : The second of the two consecutive images (from the Saarland University Computer Vision and Multimodal Computing group dataset http://www.d2.mpi-inf.mpg.de/) used in this example is shown in Fig. 4.12a. As with the example of Fig. 4.10, the foreground movement (the person's image motion) is of large magnitude, expectedly causing the detected motion region to contain the image region of the moving person in both images. The scheme has been accurate in delineating this "compound" foreground.

Fig. 4.12 Boundary-based image differencing detection via the minimization of the last two terms of Eq. (4.61): **a** the second of the two consecutive images used and, **b** the boundary of the detected region superimposed on the first image. The detected motion region includes, expectedly so, both regions covered by the object in the first and second image

4.5 Optical Flow Based Motion Detection

When optical flow is available, its norm can be used to write region-based and geodesic formulations just as with image differencing. The optical flow version of Eq. (4.41) is

$$\mathscr{E}(\gamma) = \int_{R_\gamma^c} \|W\| d\mathbf{x} + \lambda \int_{R_\gamma} d\mathbf{x}. \tag{4.62}$$

Coefficient λ acts as a threshold on the norm of optical flow. The assumption is that this norm is larger than λ for the foreground points. One can add a curve length term, $\mu \int_\gamma ds$, where μ is a positive weighing coefficient, to the functional for smooth foreground boundary, in which case the curve evolution equation is:

$$\frac{\partial \gamma}{\partial \tau} = -(\lambda - \|W\| + \mu\kappa)\,\mathbf{n}, \tag{4.63}$$

with the corresponding level set equation:

$$\frac{\partial \phi}{\partial \tau} = -(\lambda - \|W\| + \mu\kappa)\,\|\nabla\phi\|. \tag{4.64}$$

Example : *Small-extent motion*: This experiment uses two consecutive images of the marbled blocks sequence (*Marmor-2* sequence from the KOGS/ IAKS laboratory database, University of Karlsruhe, Germany). The rightmost block moves away to the left and the small center block forward to the left. The camera and the leftmost block are static. The images have been noised. The texture variation is weak at the top edges of the blocks and depth varies sharply at the blocks boundaries not in contact with the floor. The motion between the two consecutive views used is small so as to fit approximately the basic assumptions of the Horn and Schunck optical flow constraint. A vector rendering of the input (i.e., pre-computed) image motion (Chap. 3) is shown in Fig. 4.13a superimposed on the first of the two images used. The active curve boundary of the detected foreground by the application of the evolution equation (4.63) (via level set equation 4.64) is displayed in Fig. 4.13b. The active curve has correctly outlined the two moving blocks.

Example : *Large-extent motion*: The purpose of this second example is to see what kind of behavior the motion detection scheme described by Eq. (4.63) (via level set equation 4.64) has when the motion between views is significantly large. If the motion estimation succeeds in capturing this large object motion, then one would expect detection to determine a foreground which includes both the region in the first image uncovered by motion as well as the region in the second image covered by motion. Motion will, necessarily, be computed using a multiresolution scheme (Chap. 3). The two images used in this experiment are shown in Fig. 4.14a, b (the same images as in the example of Fig. 4.12, from the Saarland University Computer Vision and Multimodal Computing group dataset http://www.d2.mpi-inf.mpg.de/). We can

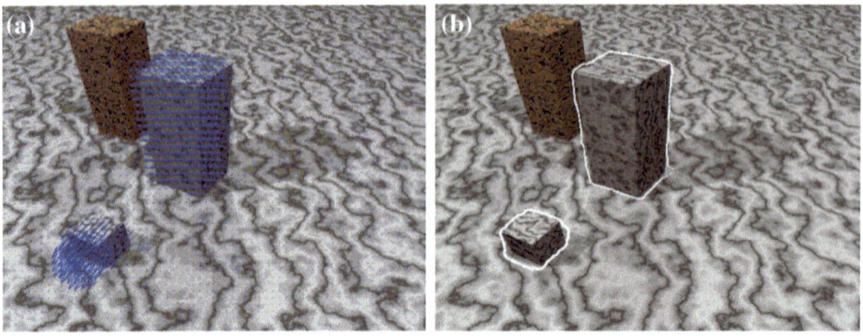

Fig. 4.13 **a** The first of the two *Marbled blocks* images used in this experiment and the input (precomputed) optical flow superimposed; **b** Application of the optical flow, region-based active curve method via Eq. (4.64): display of the boundary of the detected foreground contour superimposed on the first image

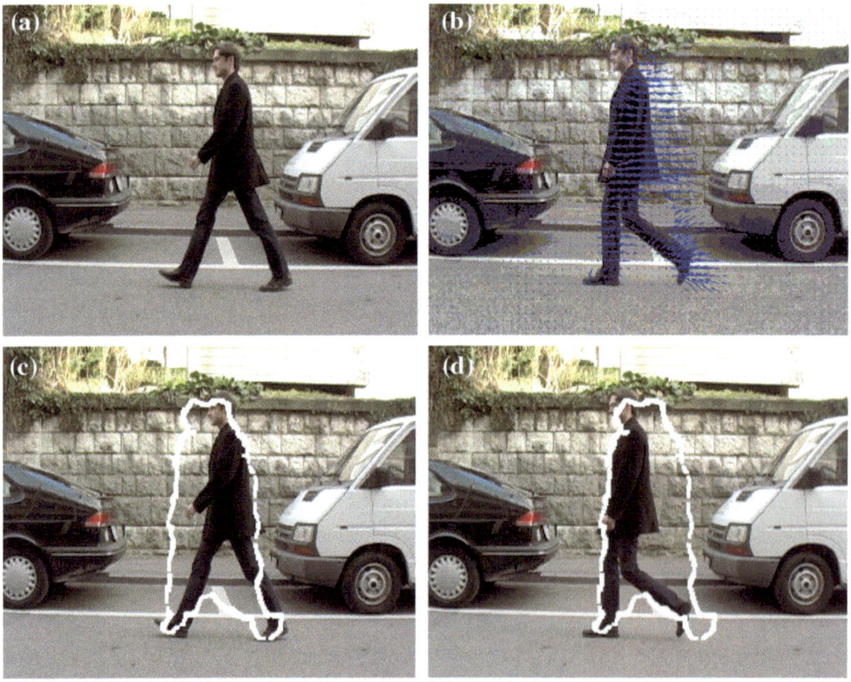

Fig. 4.14 **a** The first of the two images used to compute optical flow; **b** the second image used and the flow estimated by the Horn and Schunck algorithm embedded in multiresolution processing. The flow occurs, predictably, in both the region uncovered by motion in the first image and the region covered in the second image. Multiresolution computations have been able to capture well the overall pattern of the person's movement; **c** and **d** display the detected foreground contour, superimposed on the first and second image, respectively, showing that, as suspected, this foreground includes both the region covered and the region uncovered by motion

see that image motion between these images is quite large. The flow computed by the Horn and Shunck method embedded in multiresolution processing is superimposed on the second image. In spite of its large extent, motion was estimated so as to capture the movement of the person against the unmoving background, and one would expect this to serve well detection. This is indeed the case as shown in Fig. 4.14c, d which display the detected foreground contour (superimposed on both the first and second image). As suspected, this foreground includes both the region covered and the region uncovered by motion.

For accrued effectiveness, the study in [14] also included the image gradient geodesic of [39]:

$$\mathscr{E}(\gamma) = \int_{R_\gamma^c} \|W\| d\mathbf{x} + \lambda \int_{R_\gamma} d\mathbf{x} + \mu \int_\gamma ds + \nu \int_\gamma g(\|\nabla I\|) ds, \tag{4.65}$$

where ν is another positive weighing coefficient. The minimization of Eq. (4.65) would look for a smooth motion boundary which exhibits high image contrast to delineate a foreground of high motion. The corresponding Euler-Lagrange descent equations for the evolution of active curve γ are:

$$\frac{\partial \gamma}{\partial \tau} = -\left(\lambda - \|W\| + \mu\kappa + \kappa g\left(\|\nabla I\|\right) - < \nabla g\left(\|\nabla I\|\right), \mathbf{n} >\right) \mathbf{n}. \tag{4.66}$$

The corresponding level set evolution equation is

$$\frac{\partial \phi}{\partial \tau} = -\left(\lambda - \|W\| + \mu\kappa + \kappa g\left(\|\nabla I\|\right)\right) \|\nabla \phi\| + < \nabla g\left(\|\nabla I\|\right), \nabla \phi >. \tag{4.67}$$

If the motion field W is available, it can simply be used as data in the evolution equations. Alternatively, terms to estimate optical flow, those of the Horn and Schunck functional, for instance, can be included in the detection functional for concurrent optical flow estimation and motion detection. However, there is no compelling reason for doing so. It may be more convenient to estimate motion prior to detection by the Horn and Schunck algorithm [43, 46] or by a boundary preserving scheme such as the one in [47, 48]. By examining the basic functional Eq. (4.62), one can reasonably expect motion detection not to be dependent on highly accurate image motion because the only relevant information that is ultimately needed about the optical flow is whether the magnitude of its norm is below or above the threshold λ. However, preserving motion boundaries can be beneficial because it affords more accurate moving object boundaries.

Motion detection can also be driven by optical flow at motion boundaries. For instance, an optical flow variant of the geodesic functional Eq. (4.53) is:

$$\mathscr{F} = \int_\gamma g\left(\|W\|\right) ds, \tag{4.68}$$

Starting with an initial curve which contains the desired moving objects, a minimization of this functional with respect to curve γ will move the curve and bring it to coincide with a high optical flow amplitude boundary.

Alternatively, and more robustly, one can assume that motion boundaries are characterized not by high image motion amplitude but by high motion contrast, and minimize the geodesic functional:

$$\mathscr{F} = \int_{\gamma} g\left(\|\nabla W\|\right) ds, \tag{4.69}$$

where ∇W is the Jacobian matrix of W:

$$\nabla W = \begin{pmatrix} \frac{\partial u}{\partial x} & \frac{\partial u}{\partial y} \\ \frac{\partial v}{\partial x} & \frac{\partial v}{\partial y} \end{pmatrix} \tag{4.70}$$

and $\|.\|$ designates a matrix norm. For instance, using the Frobenius matrix norm and $g(z) = e^{-z^2}$, the curve evolution equation can be developed as:

$$\frac{\partial \gamma}{\partial \tau} = -\left(\kappa g + \langle g_1 \nabla g_2, \mathbf{n}\rangle + \langle g_2 \nabla g_1, \mathbf{n}\rangle\right) \mathbf{n}, \tag{4.71}$$

where

$$\begin{aligned} g_1\left(\gamma\left(s\right)\right) &= e^{-\|\nabla u(\gamma(s))\|^2} \\ g_2\left(\gamma\left(s\right)\right) &= e^{-\|\nabla v(\gamma(s))\|^2} \end{aligned} \tag{4.72}$$

Functions g_1 and g_2 in Eq. (4.72) above are proportional to the two-dimensional Gaussian function with zero mean and unit-variance diagonal covariance matrix, evaluated for the partial derivative vectors (u_x, u_y) and (v_x, v_y) of the optical flow component functions. Therefore, more generality, and a more flexible moving object boundary description thereof, can be achieved using a more general covariance matrix. For instance, variance parameters K_u and K_v can be used and have g_1, g_2 defined as:

$$\begin{aligned} g_1\left(\gamma\left(s\right)\right) &= e^{\frac{-\|\nabla u(\gamma(s))\|^2}{K_u^2}} \\ g_2\left(\gamma\left(s\right)\right) &= e^{\frac{-\|\nabla v(\gamma(s))\|^2}{K_v^2}} \end{aligned} \tag{4.73}$$

Example : Figure 4.15 shows an example of motion detection by the geodesic evolution equation (4.71) (*Hamburg taxi* sequence from Karlsruhe University, Institut für Algorithmen und Kognitive Systeme, http://i21www.ira.uka.de/image_sequences/). There is a single moving object in this sequence, a car which proceeds through an intersection. The first of the two consecutive images used is displayed with the initial position of the active curve in Fig. 4.15a. The final position of the curve is shown in Fig. 4.15b. The car has been correctly delineated. However, note that the curve has advanced past the car boundary through the blurred shade at the rear.

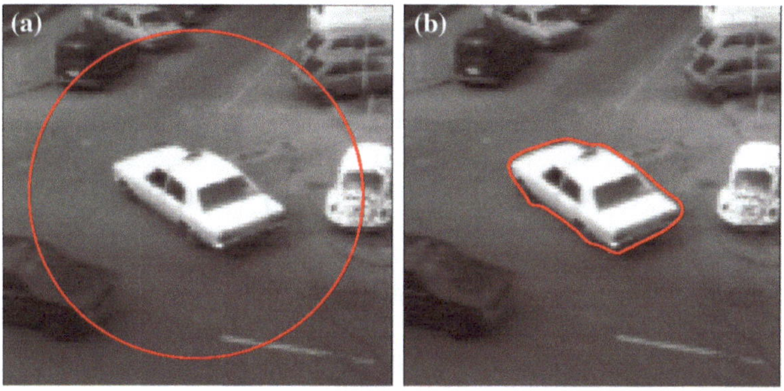

Fig. 4.15 *Hamburg Taxi* sequence: Optical flow based motion detection by minimizing the geodesic functional of Eq. (4.69) via the curve evolution equation (4.71). The white car proceeding through the intersection is the single moving object: **a** shows the image with the initial geodesic curve placed wide enough to contain the image of the moving car; **b** shows the motion boundary detected by the geodesic curve at convergence

4.6 Motion Detection with a Moving Viewing System

When the viewing system is allowed to move, region-based motion detection requires that the image motion induced by the viewing system be subtracted so that only the motion of environmental objects remains. The background becomes, ideally, motion free. However, the use of a motion-based geodesic can do without motion subtraction to correct for the viewing system movement. We will describe examples of both approaches.

4.6.1 Region Based Detection Normal Component Residuals

The study in [49] detected moving objects by simultaneously compensating for the image motion caused by the viewing system movement. The formulation, cast in a Bayesian framework, is as follows:

The basic assumption is that the background motion due to the viewing system movement can be fully characterized by a parameter vector $\boldsymbol{\theta}$. As before, let R be the region representing the foreground of moving objects and $\mathscr{R}$ the partition of Ω into R and R^c. The MAP estimate of $(R, \boldsymbol{\theta})$ is:

$$\begin{aligned}
(\hat{R}, \hat{\boldsymbol{\theta}}) &= \arg\max_{R,\boldsymbol{\theta}} P((\mathscr{R}, \boldsymbol{\theta})|m) \\
&= \arg\max_{R,\boldsymbol{\theta}} \frac{P(m|(\mathscr{R}, \boldsymbol{\theta}))P(\mathscr{R}, \boldsymbol{\theta})}{P(m)}
\end{aligned}$$

where m is a motion measurement defined on $\Omega = R \cup R^c$. The denominator, $P(m)$, is independent of $\boldsymbol{\theta}$ and R and can be removed from the arg max expression. $P(m|(\mathscr{R}, \boldsymbol{\theta})$ is the observation data term, and $P(\mathscr{R}, \boldsymbol{\theta})$ is the *a priori* term.

Assuming conditional independence of the motion measurement for $\mathbf{x} \neq \mathbf{y}$ gives:

$$P(m|(\mathscr{R}, \boldsymbol{\theta})) = \prod_{\mathbf{x} \in R} P(m(\mathbf{x})|(\mathscr{R}, \boldsymbol{\theta})) \prod_{\mathbf{y} \in R^c} P(m(\mathbf{y})|(\mathscr{R}, \boldsymbol{\theta})) \tag{4.74}$$

Therefore, the problem is equivalent to minimizing the following functional:

$$\begin{aligned} \mathscr{E}(R, \boldsymbol{\theta}) = & - \textstyle\int_R \log P(m(\mathbf{x})(\mathscr{R}, \boldsymbol{\theta}))d\mathbf{x} \\ & - \textstyle\int_{R^c} \log P(m(\mathbf{x})|(\mathscr{R}, \boldsymbol{\theta}))d\mathbf{x} \\ & - \log P(\mathscr{R}, \boldsymbol{\theta}) \end{aligned} \tag{4.75}$$

The first two terms on the righthand side of Eq. (4.75) will be specified by the observation model, or data model, and the last term by the model of prior.

Let measurement m be the normal component $W^\perp$ of optical flow:

$$W^\perp = \begin{cases} \frac{-I_t}{\|\nabla I\|} & \text{for } \|\nabla I\| \neq 0 \\ 0 & \text{for } \|\nabla I\| = 0, \end{cases} \tag{4.76}$$

Component $W^\perp$ is a data dependent measurement of motion activity which has been useful in other studies [50]. Let the optical flow *normal component residual* be:

$$W_*^\perp = W^\perp - W_c^\perp \tag{4.77}$$

where ${W_c}^\perp$ is the normal component of the image motion due to camera motion. Residual $W_*^\perp$ is a function of $\boldsymbol{\theta}$, the parameters of the image motion due to the viewing system motion. In ideal, noiseless situations, the residuals $W_*^\perp$ are zero at background points, and typically non-zero in the foreground of the moving objects.

The following data model, similar to the image differencing model of Sect. 4.4.2 but based on the normal component residual function, is a legitimate model here:

$$P(m(\mathbf{x})|(\mathscr{R}, \boldsymbol{\theta})) \propto \begin{cases} e^{-\alpha e^{-(W_*^\perp(\boldsymbol{\theta}))^2}} & \text{for } \mathbf{x} \in R \\ e^{-\beta (W_*^\perp(\boldsymbol{\theta}))^2} & \text{for } \mathbf{x} \in R^c, \end{cases} \tag{4.78}$$

where α and β are positive real constants and $\propto$ is the proportional-to symbol. This model choice will favor partitions $\mathscr{R} = R \cup R^c$ where the points in the background, R^c, have $W_*^\perp \approx 0$ and the points in the foreground R have $|W_*^\perp| \gg 0$. As a result, the formulation will look for a partition where $P(m(\mathbf{x})|(\mathscr{R}, \boldsymbol{\theta}))$ is high everywhere, i.e., in both R and R^c, the regions being described by different data models. These models will bias motion detection toward a partition where the foreground and the background display the largest possible difference in residual motion, high residuals occurring in the foreground of moving objects and low residuals in its complement. For smooth foreground boundary, and to remove small, noisy partition fragments, the following regularization term independent of $\boldsymbol{\theta}$ can be used, for some positive scalar λ:

$$P(\mathscr{R}, \boldsymbol{\theta})) \propto e^{-\lambda \int_{\partial R} ds}. \tag{4.79}$$

Let $\gamma(s) : [0, 1] \to \Omega$ be a closed simple parametric plane curve to represent the boundary ∂R of R, where s is arc length and $R = R_\gamma$ is the interior of γ. Maximizing the a posteriori probability $P((\mathscr{R}, \boldsymbol{\theta})|m)$ is equivalent to minimizing the following energy functional:

$$\mathscr{E}(R, \boldsymbol{\theta}) = \alpha \int_{R_\gamma} e^{-(W_*^\perp(\boldsymbol{\theta}))^2} d\mathbf{x} + \beta \int_{R_\gamma^c} (W_*^\perp(\boldsymbol{\theta}))^2 d\mathbf{x} + \lambda \int_\gamma ds \tag{4.80}$$

Assuming that the viewing system induced image motion is a translation over Ω, we have $\boldsymbol{\theta} = (a, b)$, where a and b are, therefore, the horizontal and vertical components of this motion, respectively. In this case,

$$W_*^\perp = -\frac{I_x a + I_y b + I_t}{\|\nabla I\|}$$

Let γ be embedded in a one-parameter family of curves $\gamma(s, \tau) : [0, 1] \times \mathbb{R}^+ \to \Omega$. The descent parameter update equations to minimize $\mathscr{E}(R, \boldsymbol{\theta}) = \mathscr{E}(\gamma, \boldsymbol{\theta})$ with respect to parameters a and b are:

$$\begin{cases} \frac{\partial a}{\partial \tau} = \alpha \int_{R_\gamma} \frac{2I_x}{\|\nabla I\|} \left(\frac{I_t + aI_x + bI_y}{\|\nabla I\|}\right) e^{-\left(\frac{I_t + aI_x + bI_y}{\|\nabla I\|}\right)^2} d\mathbf{x} \\ \quad -\beta \int_{R_\gamma^c} \frac{2I_x}{\|\nabla I\|} \left(\frac{I_t + aI_x + bI_y}{\|\nabla I\|}\right) d\mathbf{x} \\ \frac{\partial b}{\partial \tau} = \alpha \int_{R_\gamma} \frac{2I_y}{\|\nabla I\|} \left(\frac{I_t + aI_x + bI_y}{\|\nabla I\|}\right) e^{-\left(\frac{I_t + aI_x + bI_y}{\|\nabla I\|}\right)^2} d\mathbf{x} \\ \quad -\beta \int_{R_\gamma^c} \frac{2I_y}{\|\nabla I\|} \left(\frac{I_t + aI_x + bI_y}{\|\nabla I\|}\right) d\mathbf{x} \end{cases} \tag{4.81}$$

Assuming that a, b are independent of R, or γ, i.e., with a and b fixed, the Euler-Lagrange curve evolution equation to minimize $\mathscr{E}(\gamma, \boldsymbol{\theta})$ with respect to γ is given by:

$$\frac{\partial \gamma}{\partial \tau} = -(2\lambda\kappa + \alpha e^{-(W_*^\perp(a,b))^2} - \beta(W_*^\perp(a, b))^2)\mathbf{n}, \tag{4.82}$$

where $\mathbf{n}$ is the outward unit normal to γ, and κ is its mean curvature. The level set evolution equation corresponding to the evolution of γ in Eq. (4.82) is:

$$\frac{\partial \phi}{\partial \tau} = -(2\lambda\kappa + \alpha e^{-(W_*^\perp(a,b))^2} - \beta(W_*^\perp(a, b))^2))\|\nabla \phi\| \tag{4.83}$$

Fig. 4.16 *Walker* sequence: Motion detection by optical flow normal component residuals [49]: The camera tilts down slightly to cause an upward, approximately translational image motion. Images **a**, **b**, and **c** show the detected motion region in three distinct frames. The foot on the ground and part of the leg just above it are not included in the motion region because they exhibit little or no motion. The parameters used are $\alpha = 1$; $\beta = 10$; $\lambda = 5$

The algorithm can be summarized as follows:

1. Initialize γ and (a, b).
2. Perform an iteration of the descent equations for a and b in Eq. (4.81).
3. Evolve the level set ϕ of γ by an iteration of the descent Eq. (4.83).
4. Return to step 2 until convergence.

The parameters a, b can be both initialized to 0 and γ can be initially placed wide so as to contain the moving objects.

Example : The algorithm has been implemented in the image space-time domain by [49]. Motion detection is done in this case by evolving a closed regular spatiotemporal surface rather than a closed simple plane curve. However, the driving concepts remain the same in both cases. The following results have been obtained with the spatiotemporal implementation. The Walker sequence shows a man walking on a sidewalk. The camera tilts slightly down causing an upward, approximately translational image motion. Results are displayed in Fig. 4.16 showing the detected foreground of the pedestrian in three different frames. Because detection is solely based on motion activity, motion boundaries do not include portions of the pedestrian that are static during the walk. For instance, the foot on the ground and part of the leg just above are not included in the foreground because they exhibit little or no motion.

4.6.2 Detection by Optical Flow Residuals

Assume that optical flow $W = (u, v)$ over Ω is available, pre-computed, for instance by one of the methods in Chap. 3. Let W_0 be the motion over Ω induced by the viewing system movement. The field W is equal to W_0 in the background and it is the sum of object induced motion and W_0 in the foreground. Assume that W_0 is constant and write $W_0 = (a, b)$. Then motion detection can be done by minimizing the following functional:

$$\mathscr{E}(\gamma) = \int_{R_\gamma^c} (I_x a + I_y b + I_t)^2 \, d\mathbf{x} + \alpha \int_{R_\gamma^c} \|W - W_0\|^2 \, d\mathbf{x} + \lambda \int_{R_\gamma} d\mathbf{x} + \beta \int_\gamma ds, \tag{4.84}$$

where γ is, as before, a closed simple plane curve to represent the foreground contour, R_γ is the interior of γ, and (I_x, I_y, I_t) is the image spatiotemporal gradient.

The first term of the functional serves the estimation of W_0 in the background represented by the complement of R_γ. The estimation conforms to the Horn and Schunck optical flow constraint. No smoothness term is necessary since the flow is assumed constant in the region. The difference $W - W_0$ appearing in the second term is the *optical flow residual*. The last three terms are as in the case of a static viewing system except that, in this case, the viewing system motion is subtracted from the motion of the moving objects. Coefficient λ multiplying the third term plays the role of a threshold on the squared norm of the optical flow residual, i.e., on the squared norm of the image motion from which the motion induced by the viewing system movement has been subtracted. When this squared norm is larger than λ at a point, it is better to assign the point to the foreground.

The minimization equations corresponding to the functional Eq. (4.84) are derived with respect to the viewing system induced optical flow parameters a and b which appear in the two integrals over R_γ^c, and also with respect to the active curve γ which concerns all four integrals. We can adopt a descent algorithm which, after initializing γ so as to contain the moving objects, consists of repeating two consecutive steps until convergence: one step considers γ fixed and minimizes with respect to the motion parameters a and b by solving the corresponding necessary conditions:

$$\begin{aligned} &\int_{R_\gamma^c} \big(I_x(I_x a + I_y b + I_t) + \alpha(a - u)\big) \, d\mathbf{x} = 0 \\ &\int_{R_\gamma^c} \big(I_y(I_x a + I_y b + I_t) + \alpha(b - v)\big) \, d\mathbf{x} = 0 \end{aligned} \tag{4.85}$$

Since the integrands are linear functions of a and b, this amounts to determining these parameters by least squares over R_γ^c. The 2×2 system to solve is:

$$\begin{aligned} a \int_{R_\gamma^c} (I_x^2 + \alpha) \, d\mathbf{x} + b \int_{R_\gamma^c} I_x I_y \, d\mathbf{x} &= - \int_{R_\gamma^c} (I_x I_t - \alpha u) \, d\mathbf{x} \\ a \int_{R_\gamma^c} I_x I_y \, d\mathbf{x} + b \int_{R_\gamma^c} \left(I_y^2 + \alpha\right) d\mathbf{x} &= - \int_{R_\gamma^c} (I_y I_t - \alpha v) \, d\mathbf{x} \end{aligned} \tag{4.86}$$

The next step considers the parameters a and b fixed and minimizes $\mathscr{E}$ with respect to γ using the corresponding Euler-Lagrange curve evolution equation:

$$\frac{\partial \gamma}{\partial \tau} = -\left(\lambda + \beta\kappa - (I_x a + I_y b + I_t)^2 - \|W - W_0\|^2\right) \mathbf{n}, \tag{4.87}$$

to which corresponds the level set evolution equation:

$$\frac{\partial \phi}{\partial \tau} = -\left(\lambda + \beta\kappa - (I_x a + I_y b + I_t)^2 - \|W - W_0\|^2\right) \|\nabla\phi\|, \tag{4.88}$$

4.6.3 Detection by a Geodesic

When the viewing system is static, there is motion contrast at moving objects boundaries because motion is zero, ideally, in the background and it is non zero in the foreground. When the viewing system moves, the motion in the background is non-zero but there still is, in general, motion contrast at moving objects contours. Therefore, the geodesic functional Eq. (4.69):

$$\mathscr{F} = \int_\gamma g\left(\|\nabla W\|\right) ds$$

can be used. Also, function h in Eq. (4.58) is still a motion boundary indicator and a term proportional to Eq. (4.59) can be added to the geodesic functional. However, functional Eq. (4.53) is no longer valid because high values of $|I_t|$ no longer characterize motion boundaries, and neither does ∇I_t.

4.6.4 A Contrario Detection by Displaced Frame Differences

Rather than the optical flow residual, one can use the *displaced frame difference* (DFD), which is the image residual after a subtraction from optical flow of the motion induced by the viewing system movement. In the ideal, noise-free case, the DFD is zero is the background, and it is non-zero and generally significant in the foreground of moving objects. This observation was the ground for the *a contrario* detection method in [17, 51]. The scheme proceeds from two independent preliminary steps. In one step, a working set $\mathscr{R}$ of meaningful regions is extracted which have high contrast isophotes as boundaries [52]. The assumption is that the moving object boundaries have high contrast against the background, which would justify looking for them among the regions of $\mathscr{R}$.

In the other preliminary step, the viewing system motion, called the dominant motion, is represented by a linear model, affine for instance,

$$W_\theta(\mathbf{x}) = \begin{pmatrix} a_1 + a_2 x + a_3 y \\ a_4 + a_5 x + a_6 y \end{pmatrix}, \tag{4.89}$$

where x, y are the coordinates of $\mathbf{x}$ and $\theta = (a_1, \ldots, a_6)$ is the vector of model parameters, and estimated by minimizing the following objective function [53]:

$$\mathscr{E}(\theta) = \sum_{\mathbf{x}} DFD_\theta(\mathbf{x}), \tag{4.90}$$

where DFD_θ is the displaced frame difference function corresponding to θ between the current two consecutive images:

$$DFD(\theta, \mathbf{x}) = I_2(\mathbf{x} + W_\theta(\mathbf{x})) - I_1(\mathbf{x}) \tag{4.91}$$

The *DFD* residual on which detection is based is then the magnitude of the *DFD* scaled by the image gradient in the first image:

$$W_\theta^* = \frac{|DFD_\theta|}{\|\nabla I_1\|} \tag{4.92}$$

Alternatively, one can look at the smaller of the two residuals computed from two successive sets of three consecutive images [17, 51].

Following the initial steps of estimating the image motion due to the viewing system movement and determining a set of relevant regions to focus the analysis on, detection proceeds according to an explicit *a contrario* argument within regions of $\mathscr{R}$. The basis of this argument is that the residuals in the background are essentially due to white noise while in the foreground they generally have higher spatially correlated values. Therefore, the probability of the event that at least k of the n points in a foreground region R have each a residual larger that a value μ will be, for properly chosen μ, very low according to the *a contrario* hypothesis that these residuals come from the background. For practical application, a bound on this binomial probability and a set of thresholds are used to approximate the number of readings in the region that are consistent with the background ("false alarms"). This number serves to decide whether a region is part of the foreground [17, 51].

Note that the objective functional Eq. (4.8) can be looked at as an *a contrario* functional because the first integral, over region R^c, is evaluated according to the distribution of the background image and the second, over the foreground region, is evaluated according to a threshold on this distribution. A similar remark can be made about functional Eqs. (4.19) and (4.41), where the background differences and image differences in the background region are assumed normally distributed with zero mean and unit variance. A similar remark can be can be made also about the optical flow based functional Eq. (4.62). However, the method of [17, 51] remains unique in its general problem statement adapted to actual application to real images, as in the following example (courtesy of Dr Patrick Bouthemy, IRISA, Rennes, France).

The left column of Fig. 4.17 displays images from road traffic scenes imaged from a helicopter. These are difficult sequences to process because of the large extent movement of the airborne camera and the lack of definite texture on the moving vehicles. In each image, the motion in the foreground of the moving vehicles is due to both the camera movement and the vehicles own motion. In the background it is due to camera motion. The estimated image motion due to camera movement for

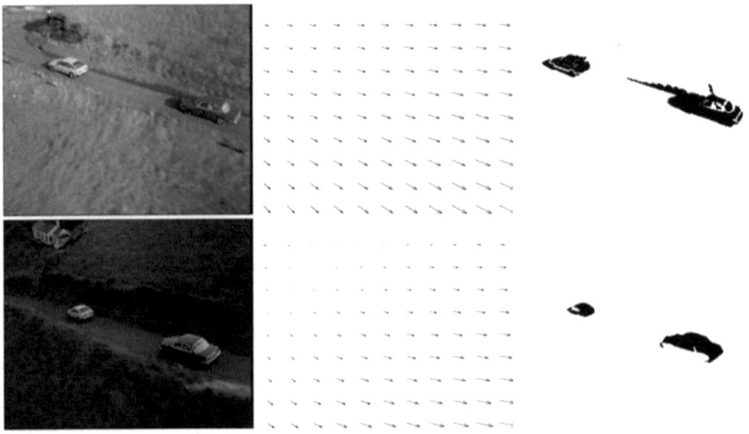

Fig. 4.17 *Left* column: images of two road traffic sequences acquired by an airborne camera. The vehicles are moving. Therefore, the motion in the foreground is due to both the camera movement and the vehicles own motion. In the background it is due to camera motion. *Middle* column: illustration of the estimated image motion due to camera movement. *Right* column: the detected motion regions by *a contrario* analysis

each sequence is illustrated graphically in the middle column. The detected motion regions are shown in the last column.

4.7 Selective Detection

The formulations of detection discussed in the preceding sections used motion to separate the image domain into a foreground of moving objects and its complement called the background. However, one may be interested not in all of the foreground objects but in a particular object or class of objects. This is selective detection. For instance, one may be interested in the human figures among the moving objects because the application is to monitor a site of human activity. There are two basic ways of determining the desired objects. One possibility is to do motion detection followed by connected component analysis to extract the moving objects individually and then perform a detailed examination of each component to determine if it fits a model description of the desired objects based on various distinguishing cues. The other possibility is to integrate a model description into the detection process. The fundamental cues are photometric descriptors, such as luminance or color, and geometric, such as the object aspect ratio or, more accurately, the object contour shape. However, motion cues describing object mobility behavior are also possible. The use of color is common in detection and general motion models have been investigated principally in motion segmentation (Sect. 4.8). There have been a few investigations of the use of object boundary shape [45, 54–61] but these generally consider classes of relatively simple shapes or relatively simple settings such as single object detection and proximity to the targeted object. The methods have not been specifically tested on important classes of objects such as humans and vehicles which are of particular interest in motion detection applications.

4.7.1 Connected Component Analysis

The distinction between the pixels of the foreground and the background is readily available in the level set implementation because the level set is of a different sign in the foreground than in the background. Therefore, a level set motion detection algorithm provides a binary image where the pixels in the foreground have a label, say 1, and those in the background a different label, say 0. A connected component analysis can then determine, by a single pass though the image, all of the connected regions of the foreground, each presumably corresponding to a different moving object. A simple algorithm is as follows:

1. Scan the image from left to right, top to bottom until reaching a row containing 1's. For every run of 1s of this row, assign a distinct label.
2. For each subsequent row:
   a. If a run of 1's is adjacent to a single run of 1s in the preceding row, assign it the label of this preceding adjacent run.
   b. If a run is adjacent to several runs in the preceding row, assign it a new (unused) label and replace the labels of all the preceding adjacent runs of 1's by this new label.

At step 2b, one does not, of course, actually go back to the runs of the preceding rows to change labels. Instead, a table of label equivalence is maintained. When the algorithm terminates (when reaching the last row of the image), all of the distinct labels point to distinct connected component objects. This algorithm was reported in [25, 62]. Before applying this algorithm, it is common to do low level "cleaning" operations such as gap filling where runs of horizontal or vertical 0's of less than a threshold length are replaced by 1's.

Human figures have been of particular interest in various motion detection applications. Several descriptive features have been considered to determine if a moving image object corresponds to a human figure. The most often included feature is the size or aspect ratio of the object bounding box. Also often used is skin colour detection within the isolated object using the (Y, C_1, C_2) color basis for better skin color representation [63]. In general, if a human figure in the image sequence has a distinguishing photometric cue, i.e., a model that shows a typical photometric appearance, say in the form of a model color histogram, the observed object photometric description is matched against this model to determine whether it is close enough to be from a human figure. Also, if the outline of the human figure, which is available via the active curve evolution algorithms described previously in this chapter, can be characterized by a few model outlines, then these models can serve to identify which isolated moving objects, obtained by connected component analysis, correspond to human figures. Finally, motion features, related to movement direction or extent, can be used if the application is such that the human figure motions of interest are distinctive in either direction or speed. Some of these features can be integrated directly in the motion detection process via some of their statistics, such as histograms, by adding a proper term in one of the motion-based objective functionals we have seen previously. This is discussed in the next section.

4.7.2 Variational Integration of Object Features in Motion Detection

We will give examples of how to integrate object distinctive cues in active curve motion detection. This can be done via feature densities. A feature density (a feature histogram in general practice) can be viewed as a marginal density of the object image or boundary it is intended to describe. As such, it can be a powerful cue when the feature is appropriately chosen. However, specific applications may involve specific cues, different from feature densities.

As alluded to in the discussion of the preceding section, cues are of three basic types, namely photometric, geometric, and motion based.

Photometric cues: Let F be a photometric feature distinctive of the image inside the region covered by the object of interest, a human figure for instance, and let $\mathcal{M}_F$ be a reference distribution of F. Model $\mathcal{M}_F$ can be learned a priori from learning examples of the desired object. Let P_F be a kernel estimate of the distribution of F inside a region R_γ. Then to account for the model description as a distinctive cue in motion detection, one can add the following properly weighed region term to the motion-based detection functional:

$$\mathcal{F}(\gamma) = \mathcal{D}(P_F, \mathcal{M}_F) \tag{4.93}$$

where $\mathcal{D}$ is a measure of the separation between distributions, such as the Bhattacharia or the Kullback-Leibler. This functional has been proposed and investigated in [26] and has been shown to be powerful enough to use for tracking. Therefore, rather than expounding on it here, as an additional term in a motion-based detection objective functional, we will defer its description until Chap. 5 on tracking for which the functional is more fitting.

Geometric cues: It is possible, via a *shape prior* [45, 54–60], to include a description of the shape of a desired object contour in a motion detection functional. In general, shape priors require an initial curve placed in the vicinity of the target object and each such prior is dedicated to the detection of a single instance in the image of the desired object.

An alternative to shape priors which can be useful when used in conjunction with a geodesic active contour has been studied in [61]. The method assumes that the desired object contour can be distinctively described against the background of other moving objects by the distribution of curvature along its delineating contour; to apply to human figures, for instance, the description requires that the human figures in the image sequence can be thus modelled.

Let $I : \Omega \subset \mathbb{R}^2 \rightarrow \mathbb{R}$ be an image function, $\gamma : [0, 1] \rightarrow \Omega$ a simple closed plane parametric curve, and $F : \Omega \subset \mathbb{R}^2 \rightarrow \mathscr{F} \subset \mathbb{R}$ a feature function from the image domain Ω to a feature space $\mathscr{F}$. Let P_γ be a kernel density estimate of the distribution of F along γ,

$$\forall f \in \mathscr{F} \quad P_\gamma(f) = \frac{\oint_\gamma K(f - F_\gamma) ds}{\mathscr{L}_\gamma}, \tag{4.94}$$

where F_γ is the restriction of F to γ, $\mathscr{L}_\gamma$ is the length of γ,

$$\mathscr{L}_\gamma = \int_\gamma ds, \tag{4.95}$$

and K is the estimation kernel. For instance, K is the Gaussian kernel of width h:

$$K(z) = \frac{1}{\sqrt{2\pi h^2}} exp^{-\frac{z^2}{2h^2}}. \tag{4.96}$$

Given a model feature distribution $\mathscr{M}$, let $\mathscr{D}(P_\gamma, \mathscr{M})$ be a similarity function between P_γ and $\mathscr{M}$. The purpose is to determine $\tilde{\gamma}$ such that

$$\tilde{\gamma} = \arg\min_\gamma \mathscr{D}(P_\gamma, \mathscr{M}). \tag{4.97}$$

To apply this formulation we need to specify the feature function, the model, the similarity function, and a scheme to conduct the objective functional minimization in Eq. (4.97).

Let $\mathscr{D}$ be the Kulback-Leibler divergence, a similarity function between distributions which has been effective in several image segmentation formulations [26, 64–66].

$$\mathscr{D}(P_\gamma, \mathscr{M}) = KL(P_\gamma, \mathscr{M}) = \int_\mathscr{F} \mathscr{M}(f) \log \frac{\mathscr{M}(f)}{P_\gamma(f)} df. \tag{4.98}$$

Higher values of the Kullback-Leibler divergence indicate smaller overlaps between the distributions and, therefore, less similarity.

Let the feature be the curvature on γ. This geometric feature is remarkable in the sense that it can be estimated from the image under the assumption that the region boundary normals coincide with the isophote normals:

$$F = \kappa_I = \operatorname{div}\left(\frac{\nabla I}{\|\nabla I\|}\right). \tag{4.99}$$

The fact that curvature along γ can be expressed as a function of the image is important in implementation because this means that it needs to be estimated only once, at the onset. However, although it is expressed in terms of the image, it remains intrinsically descriptive of the boundary geometry. Curvature, which is the rate of change of the tangent angle along the contour [67], is invariant to translation and rotation but varies with scale. However, an affine transformation of the measured values which would normalize them to map to a preset ensemble of bins, will, for all practical means, make the histogram unaffected by scale. Also, with a geometric feature such as curvature, detection can be expedited by using an edge map of the image rather than the image directly [61]. Using a working edge map will speed up processing significantly.

Let γ be embedded in a one-parameter family of curves indexed by (algorithmic) time τ:$\gamma(s, \tau) : [0, 1] \times \mathbb{R}^+ \to \Omega$, and deriving the Euler-Lagrange descent equation

$$\frac{\partial \gamma}{\partial \tau} = -\frac{\partial \mathcal{D}}{\partial \gamma}. \tag{4.100}$$

This gives [61]

$$\frac{\partial \gamma}{\partial \tau} = \left(\mathcal{G}_{KL}(P_\gamma, \mathcal{M}, F_\gamma)\kappa - \nabla \mathcal{G}_{KL}(P_\gamma, \mathcal{M}, F_\gamma) \cdot \mathbf{n}\right) \mathbf{n}, \tag{4.101}$$

where

$$\mathcal{G}_{KL}(P_\gamma, \mathcal{M}, F_\gamma) = \frac{1}{\mathcal{L}_\gamma}\left(1 - \int_{\mathcal{F}} \frac{\mathcal{M}(f)}{P_\gamma(f)} K(f - F_\gamma)\, df\right). \tag{4.102}$$

The use of $\mathcal{D}(P_\gamma, \mathcal{M})$ in conjunction with a common geodesic active contour functional [61]:

$$\mathcal{E}(\gamma, \mathcal{M}) = KL(P_\gamma, \mathcal{M}) + \lambda \int_\gamma g\left(\|\nabla I\|\right) ds \tag{4.103}$$

will seek to detect all instances of contrasted object contours in the image which are of the class of shapes described by model distribution $\mathcal{M}$. The active curve must be initialized wide out to include all the moving objects.

The behavior of Eq. (4.101) can be examined according to three cases; the first two cases assume $\mathcal{G}_{KL}$ is positive.

Case 1: The curve is in the vicinity of the target object boundary. When close to the boundary, nearly adhering to it, the curve has a feature density close to the model density, i.e., $P(F_{\gamma(\mathbf{x})}) \approx \mathcal{M}(F_{\gamma(\mathbf{x})})$ and, therefore, $\mathcal{G}_{KL} \approx 0$. As a result, the curve movement is predominantly governed by the gradient term which guides it to adhere to the desired boundary because it constrains it to move so as to coincide with local highs of the similarity between the curve feature distribution and that of the model, just as the gradient term of the common geodesic active contour drives the curve toward local highs in image gradient magnitude [39].

Case 2: The curve is not in the vicinity of the target boundary. Away from the boundary it is seeking, the active curve has, in general, a shape that is different from the model, and its feature distribution will have little overlap with the model distribution. Therefore, for most points **x** on the curve, keeping in mind that Eq. (4.101) refers to points on the curve, not on the model, we have $P(F_{\gamma(\mathbf{x})}) > \mathcal{M}(F_{\gamma(\mathbf{x})})$ and, as a result, we have $\mathcal{G}_{KL} > 0$, which means a stable evolution of γ [68].

Case 3: In the event $\mathcal{G}_{KL}$ at some point evaluates to negative at some time during curve evolution, the gradient term $\nabla\mathcal{G}_{KL} \cdot \mathbf{n}$ acts as a stabilizer of the curvature term because it constrains the curve to move along its normal to fit highs in the similarity between its feature distribution and the model distribution.

The most serious hindrance to the application of Eq. (4.103) in practice is the presence of non-targeted objects with strongly contrasted boundaries which would retain the image-based geodesic. Coefficient λ in Eq. (4.103) must be properly adjusted in such cases.

Motion cues: Motion cue integration can be handled as with photometric cues. Let F be a motion feature, the direction of motion, for instance, or the speed, and let $\mathcal{M}_F$ be the model distribution of F for the target object. To include this model in detection, a functional of the form Eq. (4.93) can be added to the objective functional.

Example : The test image is shown in Fig. 4.18a. The bottle is the target of detection. Therefore, the purpose is to detect the bottle by moving an active contour to adhere to its boundary while ignoring other boundaries in the image. The initial curve is shown in Fig. 4.18a, superimposed on the test image. Figure 4.18b shows the final position of the active contour, which coincides with the bottle boundary and only with that boundary. The feature used is curvature, computed by Eq. (4.99). Figure 4.18c displays the working edge map. The model of bottle contour on which the model histogram of curvature is learned is displayed in Fig. 4.18d. This is the outline of another bottle. Other examples can be found in [61, 69].

4.8 Motion Segmentation

Whereas motion detection partitions the image domain Ω into a foreground and a background, where the foreground corresponds to all the moving environmental objects, regardless of their motions relative to each other, motion segmentation partitions Ω into regions associated to differently moving objects. If there are N moving objects, segmentation seeks $N + 1$ regions, namely N regions $\{R_i\}_1^N$ corresponding to the moving objects and one region R^{N+1} assigned to the background, namely the complement in Ω of $\cup_1^N R_i$. A region may correspond to several distinct moving objects if these have the same motion according to the model description of this motion.

Motion-based segmentation of an image into an N−region partition can be stated as an active curve/level set functional minimization problem using $N-1$ closed regular plane curves and a model of motion to describe the motion in each region. If the motions are all described by the same parametric model, which is the case in most current studies, then the assumption is that the regions differ by the parameters of this model. Segmentation can also be done via motion detection and connected component analysis followed by an analysis of the motion within each component. This may be a more effective scheme of segmentation in applications where the number of objects is not known beforehand and the motion of objects is difficult to ascertain. This is generally the case in applications such as surveillance of site of human activity where the number of people varies in time and is not known in advance, and the human motion may be difficult to characterize by mathematically and numerically convenient descriptions such as linear parametric models [44].

There have been very few studies to investigate the problem of processing an unknown or varying number of regions in variational image segmentation. Although there are active contour schemes which include the number of regions among the

Fig. 4.18 Selective detection: **a** The bottle is the targeted object, shown with the initial active contour; **b** the contour closes on the targeted object and only on that object; **c** the working edge map; **d** the model contour which served to learn the model curvature histogram

unknowns to determine [22], either during curve evolution [70] or as a process external to curve evolution optimization [41, 71, 72], these have not been applied to motion segmentation.

Chapter 3 has reviewed parametric motion estimation and concurrent segmentation by variational active contour/level set schemes and Chap. 6 will review motion segmentation based on the movement of real objects. Although useful, these models are not always applicable. For instance, they may not be appropriate to human motion, an application where a simple linear parametric model may fail to characterize the human walk because the arms have each a different motion, as do the legs, the limbs motion being different from the motion of the rest of the body. Also the number of people in most applications may be impossible to predict and must be considered a problem variable to determine.

We will not discuss motion segmentation further; we refer the reader to Chaps. 3 and 6 of this book and to the cited literature for current variational and level set motion-based image partitioning methods, and examples of results these can produce [44, 73–80].

References

1. R. Jain, W.N. Martin, J.K. Aggarwal, Segmentation through the detection of changes due to motion. Comput. Vis. Graph. Image Process. **11**, 13–34 (1979)
2. S. Yalamanchili, J.K. Aggarwal, Segmentation through the detection of changes due to motion. Comput. Vis. Graph. Image Process. **18**, 188–201 (1981)
3. L. Wang, W. Hu, T. Tan, Recent developments in human motion analysis. Pattern Recogn. **36**(3), 585–601 (2003)
4. C. Sminchisescu, 3D human motion analysis in monocular video techniques and challenges, in *AVSS* (2006), p. 76
5. R. Poppe, Vision-based human motion analysis: an overview. Comput. Vis. Image Underst. **108**(1–2), 4–18 (2007)
6. X. Ji, H. Liu, Advances in view-invariant human motion analysis: a review. IEEE Trans. Syst. Man Cybern. Part C **40**(1), 13–24 (2010)
7. T.B. Moeslund, A. Hilton, V. Krüger, A survey of advances in vision-based human motion capture and analysis. Comput. Vis. Image Underst. **104**(2–3), 90–126 (2006)
8. Z. Sun, G. Bebis, R. Miller, On-road vehicle detection: a review. IEEE Trans. Pattern Anal. Mach. Intell. **28**(5), 694–711 (2006)
9. M. Enzweiler, D.M. Gavrila, Monocular pedestrian detection: survey and experiments. IEEE Trans. Pattern Anal. Mach. Intell. **31**(12), 2179–2195 (2009)
10. R.J. Radke, S. Andra, O. Al-Kofahi, B. Roysam, Image change detection algorithms: a systematic survey. IEEE Trans. Image Process. **14**(3), 294–307 (2005)
11. T. Bouwmans, F.E. Baf, B. Vachon, Background modelling using mixture of gaussians for foreground detection. IEEE Trans. Image Process. **1**(3), 219–237 (2008). Recent Patents on Computer Science
12. S. Jehan-Besson, M. Barlaud, G. Aubert, Detection and tracking of moving objects using a new level set based method, in *ICPR* (2000), pp. 7112–7117
13. A. Mitiche, R. Feghali, A. Mansouri, Motion tracking as spatio-temporal motion boundary detection. J. Robot. Auton. Syst. **43**, 39–50 (2003)
14. F. Ranchin, A. Chambolle, F. Dibos, Total variation minimization and graph cuts for moving objects segmentation. CoRR **abs/cs/0609100** (2006)

15. F. Ranchin, A. Chambolle, F. Dibos, Total variation minimization and graph cuts for moving objects segmentation, in *SSVM*, vol. 4485, LNCS, ed. by F. Sgallari, A. Murli, N. Paragios (Springer, Heidelberg, 2007), pp. 743–753
16. N. Paragios, R. Deriche, Geodesic active contours and level sets for the detection and tracking of moving objects. IEEE Trans. Pattern Anal. Mach. Intell. **22**(3), 266–280 (2000)
17. T. Veit, F. Cao, P. Bouthemy, An contrario decision framework for region-based motion detection. Int. J. Comput. Vis. **68**(2), 163–178 (2006)
18. T. Crivelli, P. Bouthemy, B. Cernuschi-Frías, J.-F. Yao, Simultaneous motion detection and background reconstruction with a conditional mixed-state markov random field. Int. J. Comput. Vis. **94**(3), 295–316 (2011)
19. S. Solimini, J.M. Morel, *Variational Methods in Image Segmentation* (Springer, New York, 2003)
20. G. Aubert, P. Kornpbrost, *Mathematical Problems in Image Processing: Partial Differential Equations and the Calculus of Variations* (Springer, New York, 2006)
21. S. Osher, N. Paragios, *Geometric Level Set Methods in Imaging, Vision, and Graphics* (Birkhauser, Boston, 1995)
22. A. Mitiche, I. Ben Ayed, *Variational and Level Set Methods in Image Segmentation* (Springer, New York, 2010)
23. C.R. Wren, A. Azarbayejani, T. Darrell, A. Pentland, Pfinder: real-time tracking of the human body. IEEE Trans. Pattern Anal. Mach. Intell. **19**(7), 780–785 (1997)
24. C. Stauffer, W.E.L. Grimson, Adaptive background mixture models for real-time tracking, in *CVPR* (1999), pp. 2246–2252
25. A. Rosenfeld, A. Kak, *Digital Picture Processing*, 2nd edn. (Academic, New York, 1982)
26. D. Freedman, T. Zhang, Active contours for tracking distributions. IEEE Trans. Image Process. **13**(4), 518–526 (2004)
27. S.C. Zhu, D. Mumford, Prior learning and gibbs reaction-diffusion. IEEE Trans. Pattern Anal. Mach. Intell. **19**(11), 1236–1250 (1997)
28. Y. Sheikh, M. Shah, Bayesian modeling of dynamic scenes for object detection. IEEE Trans. Pattern Anal. Mach. Intell. **27**(11), 1778–1792 (2005)
29. S. Mahamud, Comparing belief propagation and graph cuts for novelty detection, in *CVPR*, vol. 1 (2006), pp. 1154–1159
30. A. Bugeau, P. Pérez, Track and cut: Simultaneous tracking and segmentation of multiple objects with graph cuts. EURASIP J. Image Video Process. **2008**, (2008)
31. D. Mumford, J. Shah, Boundary detection by using functionals. Comput. Vis. Image Underst. **90**, 19–43 (1989)
32. Y.G. Leclerc, Constructing simple stable descriptions for image partitioning. Int. J. Comput. Vis. **3**(1), 73–102 (1989)
33. T. Chan, L. Vese, An active contour model without edges, in *International Conference on Scale-Space Theories in Computer Vision*, Greece, Corfu (1999), pp. 141–151
34. Y. Boykov, O. Veksler, R. Zabih, Fast approximate energy minimization via graph cuts. IEEE Trans. Pattern Anal. Mach. Intell. **23**(11), 1222–1239 (2001)
35. Y. Boykov, O. Veksler, Graph cuts in vision and graphics: theories and applications, in *Workshop on Mathematical Methods in Computer Vision* (2005), pp. 79–96
36. T. Chan, L. Vese, Active contours without edges. IEEE Trans. Image Process. **10**(2), 266–277 (2001)
37. O. Amadieu, E. Debreuve, M. Barlaud, G. Aubert, Inward and outward curve evolution using level set method, in *ICIP*, vol. 3 (1999), pp. 188–192
38. J.A. Sethian, *Level Set Methods and Fast Marching Methods* (Cambridge University Press, Cambridge, 1999)
39. V. Caselles, R. Kimmel, G. Sapiro, Geodesic active contours. Int. J. Comput. Vis. **22**(1), 61–79 (1997)
40. S. Kichenassamy, A. Kumar, P.J. Olver, A. Tannenbaum, A.J. Yezzi, Gradient flows and geometric active contour models, in *ICCV* (1995), pp. 810–815

41. S. Zhu, A. Yuille, Region competition: unifying snakes, region growing, and bayes/mdl for multiband image segmentation. IEEE Trans. Pattern Anal. Mach. Intell. **118**(9), 884–900 (1996)
42. M. Kass, A.P. Witkin, D. Terzopoulos, Snakes: active contour models. Int. J. Comput. Vis. **1**(4), 321–331 (1988)
43. B. Horn, B. Schunck, Determining optical flow. Artif. Intell. **17**, 185–203 (1981)
44. C. Vazquez, A. Mitiche, R. Laganiere, Joint segmentation and parametric estimation of image motion by curve evolution and level sets. IEEE Trans. Pattern Anal. Mach. Intell. **28**(5), 782–793 (2006)
45. D. Cremers, M. Rousson, R. Deriche, A review of statistical approaches to level set segmentation: integrating color, texture, motion and shape. Int. J. Comput. Vis. **62**(3), 249–265 (2007)
46. A. Mitiche, A. Mansouri, On convergence of the Horn and Schunck optical flow estimation method. IEEE Trans. Image Process. **13**(6), 848–852 (2004)
47. G. Aubert, G. Deriche, P. Kornprobst, Computing optical flow via variational thechniques. SIAM J. Appl. Math. **60**(1), 156–182 (1999)
48. R. Deriche, P. Kornprobst, G. Aubert, Optical-flow estimation while preserving its discontinuities: a variational approach, in *Asian Conference on Computer Vision* (1995), pp. 71–80
49. R. El-Feghali, A. Mitiche, Spatiotemporal motion boundary detection and motion boundary velocity estimation for tracking moving objects with a moving camera: a level sets pdes approach with concurrent camera motion compensation. IEEE Trans. Image Process. **13**(11), 1473–1490 (2004)
50. I. Cohen, G.G. Medioni, Detecting and tracking moving objects for video surveillance, in *CVPR* (1999), pp. 2319–2325
51. T. Veit, F. Cao, P. Bouthemy, Probabilistic parameter-free motion detection, in *CVPR* vol. 1 (2004), pp. 715–721
52. A. Desolneux, L. Moisan, J.-M. Morel, Edge detection by Helmholtz principle. J. Math. Imaging Vis. **14**(3), 271–284 (2001)
53. J.M. Odobez, P. Bouthemy, Robust multiresolution estimation of parametric motion models. Vis. Commun. Image Represent. **6**(4), 348–365 (1995)
54. M.E. Leventon, W.E.L. Grimson, O. Faugeras, Statistical shape influence in geodesic active contours. IEEE Conf. Comput. Vis. Pattern Recogn. **1**, 316–323 (2000)
55. Y. Chen, H. Tagare, S.R. Thiruvenkadam, F. Huang, D.C. Wilson, K.S. Gopinath, R.W. Briggs, E.A. Geiser, Using prior shapes in geometric active contours in a variational framework. Int. J. Comput. Vis. **50**(3), 315–328 (2002)
56. M. Rousson, N. Paragios, Shape priors for level set representations, in *European Conference on Computer Vision* (2002), pp. 416–418
57. A. Tsai, A.J. Yezzi, W.M.W. III, C.M. Tempany, D. Tucker, A.C. Fan, W.E.L. Grimson, A.S. Willsky, A shape-based approach to the segmentation of medical imagery using level sets. IEEE Trans. Med. Imaging **22**(2), 137–154 (2003)
58. D. Cremers, Nonlinear dynamical shape priors for level set segmentation, in *IEEE Conference on Computer Vision and, Pattern Recognition* (2007), pp. 1–7
59. D. Freedman, T. Zhang, Interactive graph cut based segmentation with shape priors. IEEE Conf. Comput. Vis. Pattern Recogn. **1**, 755–762 (2005)
60. T.F. Chan, W. Zhu, Level set based shape prior segmentation, in *IEEE Conference on Computer Vision and, Pattern Recognition* (2005), pp. 1164–1170
61. M. Ben Salah, I. Ben Ayed, A. Mitiche, Active curve recovery of region boundary patterns. IEEE Trans. Pattern Anal. Mach. Intell. **34**(5), 834–849 (2012)
62. D.H. Ballard, C.M. Brown, *Computer Vision* (Prentice Hall, New Jersey, 1982). http://homepages.inf.ed.ac.uk/rbf/BOOKS/BANDB/bandb.htm
63. Y. Dai, Y. Nakano, Face-texture model based on sgld and its application in face detection in a color scene. Pattern Recogn. **29**(6), 1007–1017 (1996)
64. A. Mansouri, A. Mitiche, Region tracking via local statistics and level set pdes, in *IEEE International Conference on Image Processing*, vol. III, Rochester, New York (2002), pp. 605–608

65. A. Myronenko, X. Song, Global active contour-based image segmentation via probability alignment, in *Computer Vision and, Pattern Recognition* (2009), pp. 2798–2804
66. F. Lecellier, S. Jehan-Besson, J. Fadili, G. Aubert, M. Revenu, Optimization of divergences within the exponential family for image segmentation, in *SSVM* (2009), pp. 137–149
67. M.P. Do Carmo, *Differential Geometry of Curves and Surfaces* (Prentice Hall, Upper Saddle River, 1976)
68. F. Guichard, J. M. Morel, *Image Analysis and PDEs* (IPAM-GBM Tutorials, 2001). http://www.ipam.ucla.edu/publications/gbm2001/gbmtut-jmorel.pdf
69. M. Ben Salah, Fonctions noyaux et a priori de forme pour la segmentation d'images et le suivi d'objets. Ph.D. dissertation, Institut national de la recherche scientifique, INRS-EMT (2011)
70. I. Ben Ayed, A. Mitiche, A region merging prior for variational level set image segmentation. IEEE Trans. Image Process. **17**(12), 2301–2313 (2008)
71. T. Kadir, M. Brady, Unsupervised non-parametric region segmentation using level sets, in *International Conference on Computer Vision* (2003), pp. 1267–1274
72. T. Brox, J. Weickert, Level set segmentation with multiple regions. IEEE Trans. Image Process. **15**(10), 3213–3218 (2006)
73. D. Cremers, C. Schnorr, Motion competition: variational integration of motion segmentation and shape regularization, in *DAGM Symposium on, Pattern Recognition* (2002), pp. 472–480
74. D. Cremers, A multiphase level set framework for motion segmentation, in *Scale Space Theories in Computer Vision*, ed. by L. Griffin, M. Lillholm, Isle of Skye, June 2003, pp. 599–614
75. A. Mansouri, J. Konrad, Multiple motion segmentation with level sets. IEEE Trans. Image Process. **12**(2), 201–220 (2003)
76. D. Cremers, S. Soatto, Motion competition: a variational approach to piecewise parametric motion segmentation. Int. J. Comput. Vis. **62**(3), 249–265 (2005)
77. T. Brox, A. Bruhn, J. Weickert, Variational motion segmentation with level sets, in *European Conference on Computer Vision*, vol. 1, (2006) pp. 471–483
78. H. Sekkati, A. Mitiche, Joint optical flow estimation, segmentation, and 3D interpretation with level sets. Comput. Vis. Image Underst. **103**(2), 89–100 (2006)
79. H. Sekkati, A. Mitiche, Concurrent 3D motion segmentation and 3D interpretation of temporal sequences of monocular images. IEEE Trans. Image Process. **15**(3), 641–653 (2006)
80. A. Mitiche, H. Sekkati, Optical flow 3D segmentation and interpretation: a variational method with active curve evolution and level sets. IEEE Trans. Pattern Anal. Mach. Intell. **28**(11), 1818–1829 (2006)

Chapter 5
Tracking

5.1 Introduction

Tracking is the process of following objects through an image sequence. In general, one may track the projected surface of an object or its outline, a patch of this surface or a set of points lying on it. Regardless of what target a tracking algorithm pursues, it must be able to characterize it so that it can seek it from one image to the next. There are three basic types of characterizations tracking can resort to: *photometric*, *geometric*, and *kinematic*. Photometric characterizations describe particular traits of the image created by the light reflected from the object, such as the distribution of color and textural properties. Color has been used often [1–5], probably because it is sensed rather than computed, in addition to being a distinctive object attribute. Texture, which refers to the spatial arrangement of image intensity giving rise to properties such as regularity, coarseness, contrast, roughness, and directionality [6], has also served tracking frequently, for instance in [7–10].

Geometric characterizations describe shape. When the purpose of tracking is to follow an image patch or an object grossly placed about a center position, it is sufficient to use a bounding figure such as a rectangle or an ellipse, as in [4, 11] for example. When tracking focuses on specific moving objects, it can use their shape when available to assist in locating them. This has been done in [12] using shape priors [13–15] with active contour representations [16].

Kinematic characterizations pertain to motion. Motion can serve tracking in two fundamental ways. First, motion can predict the position of moving objects and, therefore, assist in following them through an image sequence [17–23]. Second, motion can be used to detect moving objects and, therefore, trigger tracking. Detection of the foreground of moving objects has been investigated in Chap. 4. When tracking pursues a point set rather than a foreground region, the Harris [24], the KLT [25], and the SIFT [26] schemes have been the major methods used to detect the points.

Given an image object, which can be a point set, a patch, or a region, and its characterization, that can be photometric, geometric, or kinematic, tracking sets out

A. Mitiche and J. K. Aggarwal, *Computer Vision Analysis of Image Motion by Variational Methods*, Springer Topics in Signal Processing 10, DOI: 10.1007/978-3-319-00711-3_5,

to follow it in its course through time. With a digital image sequence, this consists of locating the object in every frame. Of course, it is not sufficient to detect the collection of moving objects in each frame because, by definition, tracking requires some form of correspondence which ties each object in one frame to the same object in the subsequent frames. This correspondence is generally embedded in the tracking process so that one does not detect a set of objects in one frame each to be matched uniquely to an object in a set of objects detected in the next frame. Instead, each moving object is followed from the current frame by positioning it in subsequent frames without having to reference other moving objects.

The earliest methods were focussed on following the trajectory of a few feature points [27–31] but, by and large, tracking has been investigated from two distinct major perspectives: *discrete-time dynamic systems theory* and *energy minimization*.

Dynamic Systems Methods

Dynamic systems methods of tracking, sometimes called data association methods [32, 33], describe the material information about a target evolution by a sequence of states $\{\mathbf{x}_k\}_k$ produced according to state transition equations $\mathbf{x}_k = \mathbf{f}_k(\{\mathbf{x}_k\}_1^{k-1}, \boldsymbol{\mu}_k)$, where $\{\boldsymbol{\mu}_k\}_k$ is an i.i.d noise sequence, describing a Markov process of order one, i.e., $\mathbf{x}_k = \mathbf{f}_k(\mathbf{x}_{k-1}, \boldsymbol{\mu}_k)$. The states $\mathbf{x}_k$ are hidden states to infer indirectly via measurements $\{\mathbf{z}_k\}_k$ to which they relate by observation equations $\mathbf{z}_k = \mathbf{h}_k(\mathbf{x}_k, \boldsymbol{\nu}_k)$, where $\{\boldsymbol{\nu}_k\}_k$ is another i.i.d noise sequence. In probabilistic form, an optimal estimate of the measurement-conditional pdf $p(\mathbf{x}_k|\{\mathbf{z}_k\}_1^k)$ which tracking aims at in the general case, can be obtained by Bayes recursive estimation. At each instant k, this consists of two steps, one of *prediction* to estimate $p(\mathbf{x}_k|\{\mathbf{z}_k\}_1^{k-1})$ using the model state-transition pdf $p(\mathbf{x}_k|\mathbf{x}_{k-1})$ and the pdf estimate at the previous instant $p(\mathbf{x}_{k-1}|\{\mathbf{z}_k\}_1^{k-1})$, followed by an *update*, in light of the new data $\mathbf{z}_k$, to estimate $p(\mathbf{x}_k|\{\mathbf{z}_k\}_1^k)$ using the likelihood $p(\mathbf{x}_k|\mathbf{z}_k)$. The process is started at an initial state whose pdf is given by the *prior* $p(\mathbf{x}_0)$.

The prediction part of these dynamic state prediction-correction iterations can be seen directly in the (Chapman-Kolmogorov) equation:

$$p(\mathbf{x}_k|\{\mathbf{z}_k\}_1^{k-1}) = \int p(\mathbf{x}_k|\mathbf{x}_{k-1})p(\mathbf{x}_{k-1}|\{\mathbf{z}_k\}_1^{k-1})d\mathbf{x}_{k-1} \tag{5.1}$$

The first term in the integrand (simplified by the first-order Markov property of the state transition pdf: $p(\mathbf{x}_k|\mathbf{x}_{k-1}, \{\mathbf{z}\}_1^{k-1}) = p(\mathbf{x}_k|\mathbf{x}_{k-1})$) of this integral expression of the prediction estimate uses the state-transition density. The second term is the previous posterior density estimate.

As to the correction part, it can been seen in the Bayes formula:

$$p(\mathbf{x}_k|\{\mathbf{z}_k\}_1^k) = \frac{p(\mathbf{z}_k|\mathbf{x}_k)p(\mathbf{x}_k|\{\mathbf{z}_k\}_1^{k-1})}{p(\mathbf{z}_k|\{\mathbf{z}_k\}_1^{k-1})}, \tag{5.2}$$

where the partition function is given by:

$$p(\mathbf{z}_k|\{\mathbf{z}_k\}_1^{k-1}) = \int p(\mathbf{z}_k|\mathbf{x}_k)p(\mathbf{x}_k|\{\mathbf{z}_k\}_1^{k-1})d\mathbf{x}_k \tag{5.3}$$

The first term in the numerator uses the observation pdf and the second term is the one computed at the prediction step. Theoretically, one can generate the desired estimates about the target from $p(\mathbf{x}_k)|\{\mathbf{z}_k\}_1^k)$, e.g., by minimum mean squared error $\tilde{\mathbf{x}}_k = E(\mathbf{x}_k|\{\mathbf{z}_k\}_1^k) = \int \mathbf{x}_k p(\mathbf{x}_k|\{\mathbf{z}_k\}_1^k)d\mathbf{x}_k$ or by maximum a posteriori criterion $\tilde{\mathbf{x}}_k = \arg\max_{\mathbf{x}_k} p(\mathbf{x}_k|\{\mathbf{z}_k\}_1^k)$. However, the optimal solution is intractable in general, and is sought only in restrictive cases. The most studied restriction is the case of Gaussian noise ($\boldsymbol{\mu}_k$'s and $\boldsymbol{\nu}_k$'s), and linear state-transition/observation equations ($\mathbf{f}_k$'s and $\mathbf{h}_k$'s), which leads to the *Kalman filter*. The formulation can be generalized to the *extended Kalman filter* [32] by a local linear expansion of the nonlinear state-transition functions about current states. Several of the earliest studies of object tracking in vision recognized the relevance of Kalman filtering in incorporating motion models and measurement uncertainties [34–41].

The assumptions underlying the Kalman filter are not applicable in general-purpose tracking and extending the filter to nonlinear state-transition and observation functions by local linear representation is fitting only to the extent that the higher-order terms are negligible. This justifies the use of practicable approximate solutions that reflect reality better. For that purpose, *particle filtering* is a general scheme that has been used for dynamic state estimation which represents the posterior pdf $p(\mathbf{x}_k|\{\mathbf{z}_k\}_1^k)$ by an appropriately sampled set of states with corresponding density values. These states/probabilities, called particles, can be used directly to estimate the most probable state or posterior mean properties $E(g(\mathbf{x})|\mathbf{z})$ relevant to an application, e.g., the mean and variance.

Essentially, particle filtering is a general method to implement nonlinear non-Gaussian Bayesian state estimation by MonteCarlo importance sampling [42]. It has been investigated under different names in various fields [43]. For instance, the *condensention* algorithm [44] used the factored sampling scheme of [45] to devise a general purpose yet simple estimation process which it applied to visual target tracking in video sequences. The condensation algorithm interprets the density multiplication in the prediction equation (5.1) as choosing with replacement a state S from the particles at the previous time step $k-1$ and then sampling from the state transition $p(\mathbf{x}_k|\mathbf{x}_{k-1} = S)$, giving a new particle S' for the current time step k. These two operations are called drift and diffusion, a borrowing from physics. Now, the density multiplication in the numerator of the Bayes/correction rule Eq. (5.2) is interpreted as using state S', which came from the drift-diffusion interpretation of the left-hand side of Eq. (5.1), to evaluate the observation density, i.e., $p(\mathbf{z}_k|\mathbf{x}_\mathbf{k} = S')$. These three consecutive operations, drift, diffusion, and evaluation, are done as many times as there are particles in the density representation to yield the current set of particles (from drift and diffusion) and an estimate of the corresponding density values (evaluation).

There have been many investigations of particle filtering variants. In particular, importance sampling has been used instead of the less efficient factored sampling of the condensation filter. A tutorial on particle filtering is offered in [43]. In general, practitioners of dynamic system tracking in vision have reported difficulties of two sorts. One difficulty relates to the objects dynamics definition. Such models generally have parameters to learn from training data while the objects are in typical motions [44]. Accurate learning may not always be practicable in general situations. Observation models may be difficult to define and to learn from data and are thus often given a tractable analytic form. Another difficulty concerns the reference density of importance sampling. It is critical that it be suited for the density being sampled in a particular application [43]. In general, a good reference density is close to the density to be sampled, and it should be easy to evaluate and simulate values from it [46].

The description of dynamic tracking above applies to a single object. When there are several objects to track, additional problems arise regarding the separation of the different dynamical objects [47, 48] , which can involve the association of measurements to current targets by multiple hypothesis evaluation and maintenance.

Energy Minimization Methods

Energy minimization methods address tracking from a quite different perspective than dynamic systems methods. The emphasis is no longer on the target dynamics but on its description so as to detect it from one instant to the next by minimizing an energy functional that embodies all of the knowledge one might have about the target, be it photometric, geometric, or kinematic. Therefore, such methods track a description, and do so by determining the image domain region in the next image which has a description that is closest to the description of the current target. This characterization gives an essential role to this region covered by the target and it is referenced explicitly in the formulation. In some methods it is a region of a typical shape, such as a rectangle or an ellipse, and in others it is of an arbitrary shape, that of the target occluding contour in the image, for instance. In the latter case, the contour can be represented by a regular closed plane curve γ which enters the problem formulation as a variable. Other variables can appear in the problem statement, for instance kinematic parameters describing the movement of the target in the interior R_γ of γ.

Let F be a feature characterizing the target. From a probabilistic standpoint, the target can be described by the distribution of this characteristic in the image domain region it covers. In this case, let q be a model density of F. This model plays the role of a prototype and can be learned a priori from training data or estimated from the feature values on the target where detected at the instant immediately preceding the current time at which we want to locate it. Let p be the density of a candidate object at the current instant. The purpose is to locate the object by minimizing a distance $d(p, q)$ between the model and candidate densities. It is clear that both the object description, via characteristic features, and the region it covers, which is where this description is evaluated, come to the forefront of the formulation. This contrasts with dynamic systems descriptions which bring the target dynamics at the foreground of processing. Also important is the choice of a distance function between the feature densities p and q or, more generally, the similarity function between the model and candidate descriptions. The choice of the similarity function often prescribes the minimization tool.

There have been several methods of tracking by energy minimization. Using ellipsoidal object regions, the study in [4] investigated a formulation which minimizes a distance between model and candidate discrete distributions of RGB color based on the Bhattacharryya coefficient. The minimization of this distance with respect to candidate objects is performed via the mean-shift algorithm [49]. Target variations in scale can be processed via pixel location normalization and variable bandwidth kernel estimate of the target density. Rather than using a fixed-shape target representation, the investigation in [50] tracks an object of arbitrary boundary represented by a regular active curve γ in the plane. The region R_γ enclosed by the curve is characterized by the distribution of a photometric feature. Locating the target at a time instant is done as in [4] by minimizing a distance between model and candidate densities. Here, however, the active contour formalism applies [51], leading to a continuous objective functional minimized by Euler-Lagrange equations via active curve evolution, and a level-set implementation for numerical efficiency [16]. Along the same vein, but using a general similarity functional rather than a distance between densities, the investigation in [52] looks for the target in the current image using regularized neighborhood matching of the target image located in the previous image. Neighborhood matching is used in a way that allows a variety of similarity functions, including probabilistic [53]. Minimization of the active curve objective functional is carried out by Euler-Lagrange equations via level sets.

Tracking by energy minimization has also been investigated from the viewpoint of motion detection. The scheme in [22], for instance, uses a geodesic active contour functional [54] with a detection term driven by temporal information, namely, the distribution of the difference at motion boundaries between consecutive images. This distribution is learned at each instant by statistical analysis of the difference image. The functional contains also a tracking term to complement detection. The tracking step, which takes over at the completion of detection, is the classical geodesic algorithm for high image gradient boundary detection. Because geodesics move in a single direction, either inward or outward, they must be placed so as to contain the target at each instant. This problem has been avoided in [55] by a region-based detection term driven by background model differencing [56, 57].

When a model of the moving object boundary shape is available, it can serves tracking. In [12] for instance, a geometric shape description of the moving object is learned from examples and embedded in a shape prior of an active curve/level set formulation. A shape prior within the active contour framework has also been used in [9] for multi-object contour tracking. Color and texture, modelled as Gaussian mixtures, assist moving object detection when there is no occlusion. The shape information, via a shape prior, is instantiated when an occlusion, detected by an external process, starts occurring.

The methods reviewed above do not use motion information about the target. However, motion can help tracking by predicting the position of a moving target, for instance according to a scheme as in [58, 59]. This can be particularly useful when occlusion, which deprives tracking of the target photometric information, starts occurring.

The remainder of this chapter offers an extended synopsis of methods which implement the fundamental driving concepts of variational tracking which we have reviewed above, namely the kernel-based formulation of [4] and mean-shift optimization which supports it (Sect. 5.2), the distribution tracking scheme of [50] (Sect. 5.3), and the temporal matching pursuit of [52] and its extensions in [53, 59] (Sect. 5.4). In one extension we will see how motion can be implicated in tracking (Sect. 5.4.3). The chapter concludes with a description of how variational tracking can be formulated as a spatiotemporal process (Sect. 5.5).

5.2 Kernel-Based Tracking

Kernel-based tracking [4] follows a standard paradigm: It uses a feature description of the target, an objective function to serve the purpose of positioning the target from one instant of observation to the next, and the minimization of this objective function. At each instant of observation, the goal is to determine the position of the target where it best fits a target model in the neighborhood of its previous position. The kernel-based tracking of [4] is characteristic in that it measures the fit of a target to the model via the Bhattacharyya distance between the feature distributions of target and model, the minimization of which reduces to the mean-shift procedure for density mode estimation, which we describe next.

5.2.1 *Mean-Shift Density Mode Estimation*

The mean shift procedure is an iterative scheme to determine a local maximum of a density function. It is developed from kernel density estimation as follows. Let $\mathbf{x}$ be a $d-$dimensional random variable with density function f. Let $V = h^d$ be the volume of the $d-$dimensional hypercube of radius h. Let $\mathscr{X} = \{\mathbf{x}_i\}_1^n$ be a set of n samples of $\mathbf{x}$. The density f can be approximated from $\mathscr{X}$ by the Parzen window estimate [60, 61] :

$$\tilde{f}(\mathbf{x}) = \frac{1}{n}\sum_{i=1}^{n}\frac{1}{V}K\left(\frac{\mathbf{x}-\mathbf{x}_i}{h}\right) \tag{5.4}$$

Function K is called a *window function* and satisfies the conditions $K(\mathbf{u}) \geq 0$ and $\int K(\mathbf{u})d\mathbf{u} = 1$ so that the estimate $\tilde{f}$ is a legitimate density function. When K is the unit hypercube window function:

$$K(\mathbf{u}) = \begin{cases} 1 & \text{for } \|\mathbf{u}_j\| < \frac{1}{2} \quad j = 1, ..., d \\ 0 & \text{otherwise,} \end{cases} \tag{5.5}$$

then the interpretation and justification of $\tilde{f}(\mathbf{x})$ is quite clear because it is the fraction of samples per unit volume falling in the window centered at $\mathbf{x}$. In the more general case, the estimate $\tilde{f}(\mathbf{x})$ is an average of functions of $\mathbf{x}$ and the samples $\{\mathbf{x}_i\}_1^n$ where K is used as an interpolation function determining the contribution of each sample $\mathbf{x}_i$ according to its distance from $\mathbf{x}$. The window function is generally called a *kernel* in computer vision applications and the window width is referred to as the *bandwidth*. The choice of the window function is not as crucial as the choice of the window width which mainly governs the estimate [62], so that the Gaussian kernel:

$$K(\mathbf{u}) = (2\pi)^{-d/2} \exp\left(-\frac{\|\mathbf{u}\|^2}{2}\right) \tag{5.6}$$

and the Epanechnikov kernel:

$$K(\mathbf{u}) = \begin{cases} \frac{1}{2V_u}(d+2)(1 - \|\mathbf{u}\|^2) & \text{for } \|\mathbf{u}\|^2 < 1 \\ 0 & \text{otherwise,} \end{cases} \tag{5.7}$$

where V_u is the volume of the unit hypersphere S_u, have been used often and without concomitant justification.

Let us take the density gradient estimate $\tilde{\nabla} f$ to be the gradient of the density estimate $\nabla\tilde{f}$. For the Epanechnikov window function, the density gradient estimate can be developed as [63]:

$$\tilde{\nabla} f(\mathbf{x}) \equiv \nabla \tilde{f}(\mathbf{x}) = \frac{1}{V} \nabla K\left(\frac{\mathbf{x} - \mathbf{x}_i}{h}\right) = \frac{n_h}{n(h^d V_u)} \frac{d+2}{h^2} \left(\frac{1}{n_h} \sum_{\mathbf{x}_i \in S_h(\mathbf{x})} (\mathbf{x}_i - \mathbf{x})\right), \tag{5.8}$$

where $S_h(\mathbf{x})$ is the hypersphere of radius h centered at $\mathbf{x}$ and n_h the number of samples in it. The term in the parenthesis on the right-hand side of Eq. (5.8) is the displacement between $\mathbf{x}$ and the average of the samples in $S_h(\mathbf{x})$ and is accordingly called the *sample mean shift*. The term $\frac{n_h}{n(h^d V_u)}$ is the Parzen window estimate $\tilde{f}$ of f when the window is the hypersphere of radius h. If the mean shift vector at $\mathbf{x}$, using the window width h, is denoted $M_h(\mathbf{x})$, then:

$$\tilde{\nabla} f(\mathbf{x}) = \alpha(\mathbf{x}) M_h(\mathbf{x}), \tag{5.9}$$

where $\alpha(\mathbf{x}) = \frac{d+2}{h^2} \tilde{f}(\mathbf{x})$. This plainly shows that this estimate of the density gradient is in the direction of the mean shift vector which, therefore, can be used to find a mode of the density given samples from it. The mean shift procedure for this purpose consists of starting at an initial position $\mathbf{y}_0$ and following the path determined by the mean shift vector, i.e., compute the path $\mathbf{y}_0, \mathbf{y}_1, \mathbf{y}_2, ...$ recursively by:

$$\mathbf{y}_{j+1} = \frac{1}{n_h} \sum_{\mathbf{x}_i \in S_h(\mathbf{y}_j)} \mathbf{x}_i, \qquad j = 0, 1, 2, ... \tag{5.10}$$

The sequence thus defined has been shown to converge [49, 64]. The scheme has been generalized to a class of circularly symmetric window functions such as the Gaussian kernel [4, 65]. For such kernels, let k be the real function defined on $[0, \infty[$ such that $K(\mathbf{u}) = k(\|\mathbf{u}\|^2)$ (called the profile of K in [4, 65]). Assuming k is differentiable, let $g = -k'$. Proceeding as with the Epanechnikov kernel one shows that the estimate of the density gradient is in the direction of the sample mean shift vector which now takes the form:

$$M_h(\mathbf{x}) = \frac{\sum_{i=1}^{n} \mathbf{x}_i g\left(\|\frac{\mathbf{x}-\mathbf{x}_i}{h}\|^2\right)}{\sum_{i=1}^{n} g\left(\|\frac{\mathbf{x}-\mathbf{x}_i}{h}\|^2\right)} - \mathbf{x}, \tag{5.11}$$

giving the recursive mean shift procedure:

$$\mathbf{y}_{j+1} = \frac{\sum_{i=1}^{n} \mathbf{x}_i g\left(\|\frac{\mathbf{y}_j-\mathbf{x}_i}{h}\|^2\right)}{\sum_{i=1}^{n} g\left(\|\frac{\mathbf{y}_j-\mathbf{x}_i}{h}\|^2\right)} \qquad j = 0, 1, 2, ... \tag{5.12}$$

Although the analysis in [4] is given in this more general case, its experimental validation suggests to use the Epanechnikov kernel rather than others. This recommendation is concordant with our earlier remark that the choice of kernel is not as crucial as that of the bandwidth [62].

5.2.2 *Tracking*

Targets are described by the distribution of a feature, color for instance, or a vector of filters response, in an elliptical image region R. Primitive geometric shapes other than ellipses may be used. Targets are compared via a similarity function of their feature densities, the Bhattacharyya coefficient, for example. Tracking seeks in the neighborhood of the previous position of the target the position that yields the largest such similarity between the target placed at that position and a model target. The initial target serves as model. This model is then updated, if necessary, while tracking progresses.

5.2.2.1 Target Description

A target is represented by the density function p of a feature $\mathbf{z}$ in the region it covers in the image. The model of the target is represented by density q. The target fit to the model can then be measured by a similarity function of the densities. For instance, one can use the negative of the Bhattacharyya coefficient ρ, as often done in image segmentation [51, 66]:

$$B(p,q) = -\rho(p,q) = -\int \sqrt{p(\mathbf{z})q(\mathbf{z})}d\mathbf{z}, \tag{5.13}$$

The study [4] suggested the following (Bhattacharyya) distance:

$$d(p,q) = \sqrt{1-\rho(p,q)}, \tag{5.14}$$

the minimization of which, one can notice, corresponds to maximizing the Bhattacharyya coefficient ρ, so that the coefficient, rather than the distance, can be used because it simplifies the analysis. In practice, $\mathbf{z}$ is quantized to m values $\mathbf{z}_1, \ldots, \mathbf{z}_m$, with corresponding probabilities $\tilde{p}_1, \ldots, \tilde{p}_m$ and $\tilde{q}_1, \ldots, \tilde{q}_m$ of the target and model. The discrete Bhattacharyya coefficient is then written as:

$$\rho(\tilde{p},\tilde{q}) = \sum_{j=1}^{m} \sqrt{\tilde{p}_j \tilde{q}_j}. \tag{5.15}$$

The m-bin histogram frequencies to estimate the target probabilities are:

$$\tilde{p}_j = \frac{1}{n} \sum_{\mathbf{x} \in R} \delta(Z(\mathbf{x}) - \mathbf{z}_j), \tag{5.16}$$

where R is the elliptical region covered by the target and $n = card(R)$, and $Z : \mathbf{x} \in D \to Z(\mathbf{x}) \in \mathcal{Z}$ is the feature value mapping on the discrete image domain D. Alternatively, one can use weighed frequencies as in [4] to give less importance to points away from the region center which are more vulnerable to occlusion:

$$\tilde{p}_j = c \sum_{\mathbf{x} \in R} K\left(\left\|\frac{\mathbf{y}-\mathbf{x}}{h}\right\|\right) \delta(Z(\mathbf{x}) - \mathbf{z}_j), \tag{5.17}$$

where $\delta(0) = 1$, and 0 otherwise, K is a window function of width h, $\mathbf{y}$ is the center of R, and c is the normalizing constant so that the frequencies sum to 1. The target model, centered at a fixed location, has similarly expressed density estimates.

5.2.2.2 Mean-Shift Tracking

A linear expansion of the Bhattacharyya coefficient in the neighborhood of $\mathbf{y}_0$ gives:

$$\rho(\tilde{p}(\mathbf{y}),\tilde{q}) = \sum_{j=1}^{m} \sqrt{\tilde{p}_j(\mathbf{y})\tilde{q}_j} \approx \frac{1}{2}\rho(\tilde{p}(\mathbf{y}_0),\tilde{q}) + \frac{1}{2}\sum_{j=1}^{m} \tilde{p}_j(\mathbf{y})\sqrt{\frac{\tilde{q}_j}{\tilde{p}_j(\mathbf{y}_0)}} \tag{5.18}$$

The maximization with respect to $\mathbf{y}$ of the sum on the righthand side of the $\approx$ sign in Eq. (5.18) involves only the second term which, using Eq. (5.17), can be written:

$$\frac{c}{2}\sum_{\mathbf{x}\in R}\alpha(\mathbf{x})K\left(\left\|\frac{\mathbf{y}-\mathbf{x}}{h}\right\|\right), \tag{5.19}$$

where the coefficients $\alpha(\mathbf{x})$ are given by:

$$\alpha(\mathbf{x}) = \sum_{j=1}^{m}\sqrt{\frac{\tilde{q}_j}{\tilde{p}_j(\mathbf{y}_0)}}\cdot\delta(Z(\mathbf{x})-\mathbf{z}_j) \tag{5.20}$$

We see that expression Eq. (5.19) resembles the righthand side of the Parzen window estimate in Eq. (5.4), although the window function K in Eq. (5.19) refers to image positions rather than random variables. The coefficients being independent of the running variable $\mathbf{y}$, the sample mean shift analysis remains valid and, therefore, the mean shift procedure can be applied to Eq. (5.19) by computing the sequence:

$$\mathbf{y}_{j+1} = \frac{\sum_{\mathbf{x}\in R}\mathbf{x}\alpha(\mathbf{x})g\left(\left\|\frac{\mathbf{y}_j-\mathbf{x}}{h}\right\|^2\right)}{\sum_{\mathbf{x}\in R}\alpha(\mathbf{x})g\left(\left\|\frac{\mathbf{y}_j-\mathbf{x}}{h}\right\|^2\right)} \qquad j = 0, 1, 2, ... \tag{5.21}$$

In practice, one must check that the sequence is properly stepped to converge. A scheme must also be used for scale adaption when the target moves in depth. These and other details are explained in [4].

Example: There are several examples of the kernel-based mean-shift tracking in [4]. Figure 5.1 (Courtesy of D. Comaniciu, reproduced with IEEE permission) shows the results with the *Mug* sequence of one of the examples. The object to track is a mug using an ellipsoid. The RGB color is the feature to describe the target, with values quantized into $16 \times 16 \times 16$ bins. The Epanechnikov profile is used for histogram computations and the mean shift iterations are based on weighted averages. The sequence is about 1,000 frames long. The target model is the image of a mug in one of the frames (frame 60). The algorithm was tested with fast motion (frame 150), abrupt appearance variation (frame 270), target rotation (frame 270), and scale change (frames 360–960). Tracking has kept the ellipsoid with the target until the end of the sequence.

5.3 Active Curve Density Tracking

In kernel-based tracking which we examined in the previous section, the target had a fixed primitive shape modulo scale. The target in [4] was an elliptical region. At each instant of observation the object of tracking was to locate a region of the same shape in which the density of a feature closely matches a model density. This matching minimized the Bhattacharyya distance between model and target region densities and the minimization was done via the mean-shift procedure thanks to a window function

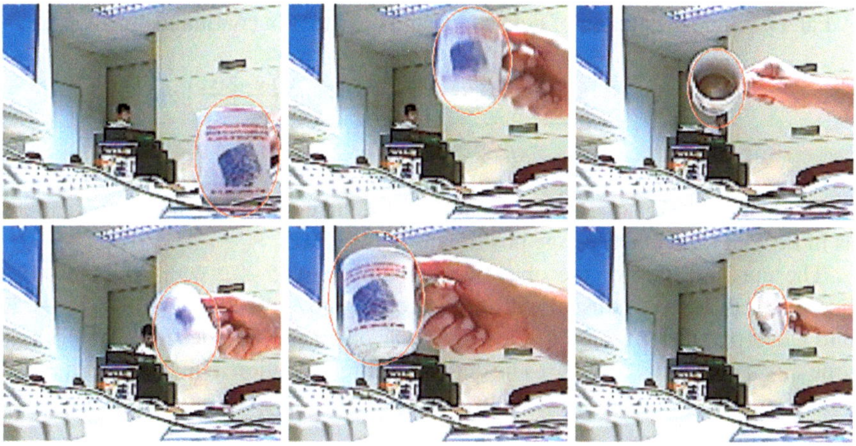

Fig. 5.1 Kernel/mean shift tracking [4]: results are shown for frames 60, 150, 240, 270, 360, 960

weighed expression of the target feature histogram (Eq. 5.17). The method we will now examine [50, 67] also uses a feature density description of model and target, and the paradigm of tracking by locating at each instant of observation the region in which the feature density is most similar to the model density. However, there are two major differences in that the target region is no longer constrained to have a fixed shape and tracking is done by an active curve which moves to coincide with the target boundary. This active curve is a variable in the formulation and its movement is governed by the Euler-Lagrange equations corresponding to the minimization of the model and target density similarity.

Let $I : \mathbf{x}, t \in \Omega \times T \rightarrow I(\mathbf{x}, t)$ be an image sequence, where Ω is the image domain and T the sequence duration interval. Let $\mathbf{z} \in \mathscr{Z}$ be a descriptive image feature of the target and $Z : \mathbf{x} \in \Omega \rightarrow Z(\mathbf{x}) \in \mathscr{Z}$ the feature function. As with other active curve formulations in other chapters of this book, let $\gamma(s) : [0, 1] \rightarrow \Omega$ be a closed simple parametric curve of the plane and R_γ its interior. Finally, let p, q be the densities of $\mathbf{z}$ for the target and the model, respectively.

5.3.1 The Bhattacharyya Flow

To determine a region $R \in \Omega$ in which the density of $\mathbf{z}$ resembles most the model density, one determines the curve γ which maximizes the Bhattacharyya coefficient ρ of p and q, p being a function of γ. The Bhattacharyya coefficient ρ is written, in this context of active curve representation of target boundaries:

$$\rho(p, q) = \rho(p(\gamma), q)) = \int_{\mathscr{Z}} \sqrt{p(\mathbf{z}, \gamma)q(\mathbf{z})} d\mathbf{z} \tag{5.22}$$

A Parzen window estimate of density p can be written in continuous form as:

$$p(\mathbf{z}, \gamma) = \frac{\int_{R_\gamma} K\left(\frac{\mathbf{z}-Z(\mathbf{x})}{h}\right) d\mathbf{x}}{\int_{R_\gamma} d\mathbf{x}} = \frac{\int_{R_\gamma} K\left(\frac{\mathbf{z}-Z(\mathbf{x})}{h}\right) d\mathbf{x}}{A(R_\gamma)}, \tag{5.23}$$

$A(R_\gamma)$ being the area of R_γ. Note that kernel K here applies to the feature variable as in Eq. (5.4) and not to image position as in Eq. (5.17).

The maximization of the objective functional Eq. (5.22), with p expressed in Eq. (5.23), is done as is usual with active contours by embedding $\gamma : [0, 1] \rightarrow \mathbb{R}^2$ into a one-parameter family of closed regular plane curves indexed by algorithmic time τ: $\gamma : [0, 1] \times \mathbb{R}^+ \rightarrow \mathbb{R}^2$. The curve evolution equation to maximize Eq. (5.22) with respect to γ is then given by the functional derivative of ρ with respect to γ:

$$\frac{\partial \gamma}{\partial \tau} = \frac{\partial \rho}{\partial \gamma}, \tag{5.24}$$

which is derived as follows:

$$\frac{\partial \rho}{\partial \gamma} = \frac{1}{2} \int_{\mathcal{Z}} \sqrt{\frac{q(\mathbf{z})}{p(\mathbf{z}, \gamma)}} \frac{\partial p(\mathbf{z}, \gamma)}{\partial \gamma} d\mathbf{z}, \tag{5.25}$$

with:

$$\begin{aligned} \frac{\partial p}{\partial \gamma}(\mathbf{z}, \gamma) &= \frac{A(R_\gamma) \frac{\partial \int_{R_\gamma} K\left(\frac{\mathbf{z}-Z(\mathbf{x})}{h}\right) d\mathbf{x}}{\partial \gamma} - \frac{\partial A(R_\gamma)}{\partial \gamma} \int_{R_\gamma} K\left(\frac{\mathbf{z}-Z(\mathbf{x})}{h}\right) d\mathbf{x}}{A(R)^2} \\ &= \frac{1}{A(R_\gamma)} \left(K\left(\frac{\mathbf{z} - Z(\gamma)}{h}\right) - p(\mathbf{z}, \gamma) \right) \mathbf{n}, \end{aligned} \tag{5.26}$$

where $\mathbf{n}$ is the outward unit normal function of γ. Substitution of Eq. (5.26) in Eq. (5.25) gives:

$$\frac{\partial \gamma}{\partial \tau} = V_\rho \mathbf{n}, \tag{5.27}$$

where the velocity V_ρ is:

$$V_\rho = \frac{1}{2A(R_\gamma)} \left(\int_{\mathcal{Z}} K\left(\frac{\mathbf{z} - Z(\gamma)}{h}\right) \sqrt{\frac{q(\mathbf{z})}{p(\mathbf{z}, \gamma)}} d\mathbf{z} - \rho(p(\gamma), q) \right). \tag{5.28}$$

When K is the Dirac delta function we can write:

$$V_\rho = \frac{1}{2A(R_\gamma)} \left(\sqrt{\frac{q(Z(\gamma))}{p(Z(\gamma))}} - \rho(p(\gamma), q) \right) \tag{5.29}$$

We will now develop the Kullback-Leibler flow equations by adopting similar manipulations. We will then follow with an intuitive interpretation of both the Bhattacharyya and the Kullback-Leibler flows.

5.3.2 The Kullback-Leibler Flow

The formulation can use other density similarity functions, the Kullback-Leibler divergence, for instance:

$$KL(p,q) = KL(p(\gamma),q) = \int_{\mathscr{Z}} q(\mathbf{z}) \log \frac{q(\mathbf{z})}{p(\mathbf{z},\gamma)} d\mathbf{z} \tag{5.30}$$

The corresponding curve evolution equation to minimize the objective functional can be written:

$$\frac{\partial \gamma}{\partial t} = -\frac{\partial KL}{\partial \gamma}, \tag{5.31}$$

with

$$\frac{\partial KL}{\partial \gamma} = -\int_{\mathscr{Z}} \frac{q(\mathbf{z})}{p(\mathbf{z},\gamma)} \frac{\partial p(\mathbf{z},\gamma)}{\partial \gamma} d\mathbf{z} \tag{5.32}$$

Substitution of Eq. (5.26) in Eq. (5.32) gives:

$$\frac{\partial \gamma}{\partial \tau} = V_{KL} \mathbf{n} \tag{5.33}$$

with:

$$V_{KL} = \frac{1}{A(R_\gamma)} \left(\int_{\mathscr{Z}} K\left(\frac{\mathbf{z} - Z(\gamma)}{h}\right) \frac{q(\mathbf{z})}{p(\mathbf{z},\gamma)} d\mathbf{z} - 1 \right). \tag{5.34}$$

When K is the Dirac function we can write:

$$V_{KL} = \frac{1}{A(R_\gamma)} \left(\frac{q(Z(\gamma))}{p(Z(\gamma))} - 1 \right). \tag{5.35}$$

Velocity Eqs. (5.29) and (5.35) hint at the behaviour of the Bhattacharyya and the Kulback-Leibler curve evolution equations. The Kulback-Leibler velocity Eq. (5.35) amounts to a likelihood ratio test to evaluate the hypothesis that the feature value at a point $\mathbf{x} \in \gamma$ is drawn from model q against the hypothesis that it is drawn from the p in R_γ. If $q(Z(\mathbf{x})) > p(Z(\mathbf{x}))$, the likelihood ratio $\frac{q(Z(\mathbf{x}))}{p(Z(\mathbf{x}))}$ is greater than 1, causing the velocity to be positive. Therefore, the movement of the curve at $\mathbf{x}$ will be in the direction of its outward unit normal and it will expand to include $\mathbf{x}$ in R_γ, thereby drawing p closer to q at $Z(\mathbf{x})$. If, instead, $q(Z(\mathbf{x})) \le p(Z(\mathbf{x}))$, the curve will retract from $\mathbf{x}$ which will, therefore, be in R_γ^c. Examining velocity Eq. (5.29) reveals a similar behaviour of the Bhattacharyya flow. The difference is that the likelihood ratio hypothesis threshold is the current Bhattacharyya coefficient at the feature value measured at **x**, $Z(\mathbf{x})$, rather that being fixed at 1 as for the Kullback-Leibler flow.

5.3.3 Level Set Equations

As with other active curve formulations in other chapters of this book, curve γ is represented implicitly as the level set zero of a function ϕ, called the level set function, $\phi : \mathbb{R}^2 \to \mathbb{R}$, i.e., γ is the set $\{\phi = 0\}$. Recall that when a curve moves according to $\frac{d\gamma}{d\tau} = V\mathbf{n}$, its level set function evolves according to [16]

$$\frac{\partial \phi}{\partial \tau}(\tau) = V \|\nabla \phi\|. \tag{5.36}$$

We also recall that by evolving ϕ rather than γ, the topological variations of the curve occur automatically and its position can be recovered as the level zero of ϕ at any time. Would the curve be evolved directly, via an explicit representation as a set of points, its topological changes would not be implementable in general. We have reviewed the level set representation in Chap. 2 and there are extensive explanations in [16] on effective numerical algorithms to implement level set evolution equations.

In our case, velocity V is given by Eq. (5.28) for the Bhachattaryya flow and by Eq. (5.34) for the Kullback-Leibler flow. i.e., the corresponding level set function evolution equations are given by Eq. (5.36), where $V = V_B$ for the Bhachattaryya flow and $V = V_{KL}$ for the Kullback-Leibler flow.

Although the curve evolution equations Eq. (5.27) and Eq. (5.33) refer to the points on γ, the velocity can be computed everywhere in Ω. Therefore, the level set equation can be generalized so as to evolve the level set function everywhere in Ω. Free implementations can also be found on the web.

Example: This example shows an application of tracking by the Kullback-Leibler flow of the distribution matching scheme of [50]. It uses a sequence showing a person walking (Fig. 5.2). The intensity profile corresponding to the person contrasts well with that of its background and is well defined. The target intensity distribution model is learned from the first frame using a manual segmentation (top image of Fig. 5.2). Using this model, tracking was able to follow well its initial target through the sequence.

The methods we described in the previous two sections used a model feature distribution to pursue the target from one instant of observation to the next. The model is normally learned off-line but could possibly be acquired from the target in the previous image. The region-based matching scheme we will now describe [52] and its extensions [53, 59] do not use a target model learned a priori, but take the image of the target in the previous frame as the reference to match in the frame of the current instant.

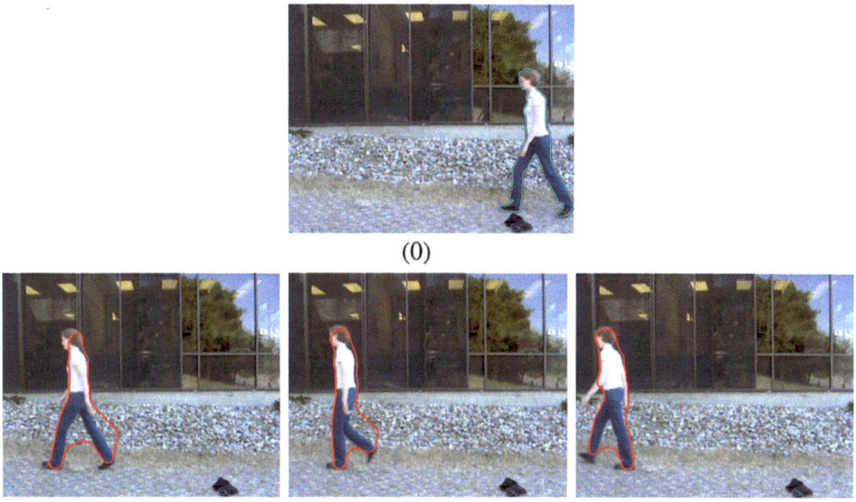

(0)

(52) (57) (64)

Fig. 5.2 Tracking distributions [50] algorithm using the Kullback-Leibler flow: the person is followed throughout the sequence by matching the intensity distribution on the target from one image to the next. The image labelled (0) is the first frame of the sequence and it displays the manual segmentation used to learn the target image statistics and start tracking. The other images are frames toward the end of the sequence (as numbered below the images) showing the results of tracking

5.4 Tracking by Region-Based Matching

Let I^{k-1} and I^k be two images with common domain Ω at two consecutive instants of observation $k-1$ and k. Let R_0 be the target region at the previous instant $k-1$ and R_1 its corresponding unknown region at the current instant k. Region R_1 is the object of tracking. The following analysis [52] is given for scalar images but can be rewritten for multivalued images such as color or vectors of filter responses.

5.4.1 *Basic Formulation*

Assume that R_0 can be mapped onto R_1 and R_0^c onto R_1^c and that the corresponding images differ by noise. Specifically, assume that there exists $\psi \in \Psi$ such that

$$I^k(\psi(\mathbf{x})) = I^{k-1}(\mathbf{x}) + \mu(\mathbf{x}) \quad \forall \mathbf{x} \in \Omega, \tag{5.37}$$

where Ψ is a set of allowable mapings and μ is stationary zero-mean Gaussian white noise of standard deviation σ. This image model does not actually apply to parts of Ω which are covered/uncovered by the target motion or occlusion. However, it is a useful model to formulate the problem so as to arrive at a practicable effective algorithm. Using this model and some mild probabilistic assumptions, the study in [52] converts the MAP estimation of R_1:

$$\tilde{R_1} = \arg \max_{R \in \Omega} P(R_1 = R | I^{k-1}, I^k, R_0) \tag{5.38}$$

into a functional minimization problem:

$$\tilde{\gamma} = \arg \min_{\gamma} \mathscr{E}(\gamma | I^{k-1}, I^k, R_0), \tag{5.39}$$

where

$$\mathscr{E}(\gamma | I^{k-1}, I^k, R_0) = \int_{R_\gamma} \xi_1(\mathbf{x}) d\mathbf{x} + \int_{R_\gamma^c} \xi_2(\mathbf{x}) d\mathbf{x} + \lambda \int_\gamma ds \tag{5.40}$$

with functions ξ_1 and ξ_2 are given by:

$$\begin{cases} \xi_1(\mathbf{x}) = \inf_{\{z: \|\mathbf{z}\| \leq \alpha, \mathbf{x}+\mathbf{z} \in R_0\}} \frac{(I^k(\mathbf{x}) - I^{k-1}(\mathbf{x}+\mathbf{z}))^2}{2\sigma^2} \\ \xi_2(\mathbf{x}) = \inf_{\{z: \|\mathbf{z}\| \leq \alpha, \mathbf{x}+\mathbf{z} \in R_0^c\}} \frac{(I^k(\mathbf{x}) - I^{k-1}(\mathbf{x}+\mathbf{z}))^2}{2\sigma^2}, \end{cases} \tag{5.41}$$

and $\gamma : s \in [0, 1] \rightarrow \gamma(s) \in \Omega$ is closed simple plane curve representation of R_1 parametrized by arc length and R_γ is its interior.

The length term, the third on the righthand side of the Eq. (5.40), imposes a smooth boundary on the solution. The image model Eq. (5.37) manifests in the *inf* argument of the objective functional data terms ξ_1, ξ_2 which result from taking $-log$ of the matching error probability according to a zero-mean Gaussian of σ^2 variance. There also is in these data terms the tacit assumption which limits to α the extent of the motion between the images at the previous and current instants $k-1$ and k. However, we will see that the formulation can gain from basic motion information such as an estimate of the target global motion (Sect. 5.4.3).

By embedding γ into a one-parameter family of closed simple plane curves indexed by algorithmic time τ: $\gamma : [0, 1] \times \mathbb{R}^+ \rightarrow \mathbb{R}^2$, the curve evolution equation to minimize Eq. (5.40) with respect to γ is given by:

$$\frac{\partial \gamma}{\partial \tau} = -(\xi_1 - \xi_2 + \lambda \kappa)\, \mathbf{n}, \tag{5.42}$$

where κ is the curvature function of γ and $\mathbf{n}$ its outward unit normal function. The evaluations are, of course done on the curve at each instant, i.e.,

$$\frac{\partial \gamma}{\partial \tau}(s, \tau) = -(\xi_1(\gamma(s, \tau)) - \xi_2(\gamma(s, \tau)) + \lambda \kappa(s, \tau))\mathbf{n}(s, \tau) \tag{5.43}$$

The corresponding level set equation, generalized to apply at all $\mathbf{x} \in \Omega$, is:

$$\frac{\partial \phi}{\partial \tau}(\mathbf{x}, \tau) = -\left(\xi_1(\mathbf{x}) - \xi_2(\mathbf{x})\right) + \lambda \kappa(\mathbf{x}, \tau))\|\nabla \phi(\mathbf{x}, \tau)\| \tag{5.44}$$

Recall that the curvature function can be written in terms of the level set function as the divergence of its unit gradient:

$$\kappa = \nabla \cdot \frac{\nabla \phi}{\|\nabla \phi\|}, \tag{5.45}$$

when ϕ is positive inside γ, negative outside, and the normal $\mathbf{n}$ of γ is oriented outward.

The basic formulation of [52] described above can be generalized by extending the definition of the data terms ξ_1, ξ_2 in Eq. (5.41) to account for local image statistics [53] and for the target shape and motion [59]. We will present these extensions in the next two sections. First, we give an illustrative example of the behavior of the model-distribution tracking scheme we have just described.

Example: This example uses the same sequence as the previous one for the model-density tracking scheme. It implements the basic region-based matching scheme of [52]. The results are shown in (Fig. 5.3). The same initialization is used in this example as in the previous. The matching scheme, aided by the good image contrast at the walker image boundary, allowed tracking to adhere to its target through the sequence. The good performance compared to the model-density matching scheme of the previous example is mainly due to the use of the previously detected target, rather than a fixed target model, as the target instance to look for in the current frame.

(47) (57) (64)

Fig. 5.3 Tracking by the basic region-based matching: This is the same sequence as in the previous example of tracking distributions of Fig. 5.2. The same initialization is used. The active contour delineates well the silhouette of the person mainly thanks to the good contrast at the walker image boundary

5.4.2 Matching Local Statistics

Let $p_{I,\mathbf{x}}$ designate the probability density function of I at $\mathbf{x}$. The data terms ξ_1, ξ_2 can be generalized so that the level set evolution equation (5.44) takes the form:

$$\begin{aligned}\frac{\partial \phi}{\partial \tau}(\mathbf{x}, \tau) = - \Big\{ & \inf_{\{z:\|\mathbf{z}\| \leq \alpha, \mathbf{x}+\mathbf{z} \in R_0\}} D\big(p_{I^k,\mathbf{x}}, p_{I^{k-1},\mathbf{x}+\mathbf{z}}\big) \\ & - \inf_{\{z:\|\mathbf{z}\| \leq \alpha, \mathbf{x}+\mathbf{z} \in R_0^c\}} D\big(p_{I^k,\mathbf{x}}, p_{I^{k-1},\mathbf{x}+\mathbf{z}}\big) \\ & + \lambda \kappa(\mathbf{x}, \tau)\Big\} \|\nabla \phi(\mathbf{x}, \tau)\|, \end{aligned} \tag{5.46}$$

where $D(p, q)$ designates a density separation function, for instance the Kullback-Leibler divergence or the Bhattacharyya coefficient which we used previously. The densities $p_{I^k,\mathbf{x}}$ and $p_{I^{k+1},\mathbf{x}+\mathbf{z}}$ appearing in Eq. (5.46) can be approximated in a neighborhood of $\mathbf{x}$ and $\mathbf{x}+\mathbf{z}$, respectively, for instance by a nonparametric kernel estimate such as a histogram, or by assuming a parametric form and estimating the parameters. These densities reflect local image statistics.

Let us for a moment put aside the length term of the functional and examine the behaviour of the level set evolution at a specific point $\mathbf{x}$ in the current image k at time τ. If the local statistics of I^k at $\mathbf{x}$ are closer to the local statistics of I^{k-1} at a point in R_0 than at a point in R_0^c, i.e.,

$$\inf_{\{z:\|\mathbf{z}\| \leq \alpha, \mathbf{x}+\mathbf{z} \in R_0\}} D\big(p_{I^k,\mathbf{x}}, p_{I^{k-1},\mathbf{x}+\mathbf{z}}\big) - \inf_{\{z:\|\mathbf{z}\| \leq \alpha, \mathbf{x}+\mathbf{z} \in R_0^c\}} D\big(p_{I^k,\mathbf{x}}, p_{I^{k-1},\mathbf{x}+\mathbf{z}}\big) \leq 0, \tag{5.47}$$

then $\frac{\partial \phi}{\partial \tau}(\mathbf{x}, \tau) \geq 0$ and ϕ monotonically increases to become eventually positive, in which case $\mathbf{x}$ is claimed by R_1 as it should, where R_1 is the estimate of the tracked region at the current instant k. If, instead, the local statistics of I^k at $\mathbf{x}$ are closer to the local statistics of I^{k-1} at a point in R_0^c than at a point in R_0, then $\frac{\partial \phi}{\partial \tau}(\mathbf{x}, \tau) \leq 0$ and point $\mathbf{x}$ will eventually go to R_1^c as it should.

As an application, consider the case when $p_{I^k,\mathbf{x}}$ is normally distributed with mean

$$\mu_{1,\mathbf{x}} = \frac{\int_{\mathscr{B}(\mathbf{x},\beta)} I^k(\mathbf{y}) d\mathbf{y}}{\int_{\mathscr{B}(\mathbf{x},\beta)} d\mathbf{y}}, \tag{5.48}$$

where $\mathscr{B}(\mathbf{x}, \beta)$ is the ball of radius β centered at $\mathbf{x}$, and variance

$$\sigma_{1,\mathbf{x}}^2 = \frac{\int_{\mathscr{B}(\mathbf{x},\beta)} (I^k(\mathbf{y}) - \mu_{1,\mathbf{x}})^2 d\mathbf{y}}{\int_{\mathscr{B}(\mathbf{x},\beta)} d\mathbf{y}}, \tag{5.49}$$

and, similarly, consider that $p_{I^{k-1},\mathbf{x}+\mathbf{z}}$ is normally distributed with mean

$$\mu_{2,\mathbf{x}+\mathbf{z}} = \frac{\int_{\mathscr{B}(\mathbf{x}+\mathbf{z},\beta)} I^{k-1}(\mathbf{y})d\mathbf{y}}{\int_{\mathscr{B}(\mathbf{x}+\mathbf{z},\beta)} d\mathbf{y}} \tag{5.50}$$

and variance

$$\sigma^2_{2,\mathbf{x}+\mathbf{z}} = \frac{\int_{\mathscr{B}(\mathbf{x}+\mathbf{z},\beta)} (I^{k-1}(\mathbf{y}) - \mu_{2,\mathbf{x}+\mathbf{z}})^2 d\mathbf{y}}{\int_{\mathscr{B}(\mathbf{x}+\mathbf{z},\beta)} d\mathbf{y}}, \tag{5.51}$$

then the Kullback-Leibler divergence between $p_{I^k,\mathbf{x}}$ and $p_{I^{k-1},\mathbf{x}+\mathbf{z}}$ is given by:

$$KL(p_{I^k,\mathbf{x}}, p_{I^{k-1},\mathbf{x}+\mathbf{z}}) = \frac{1}{2}\Big(\log \frac{\sigma^2_{2,\mathbf{x}+\mathbf{z}}}{\sigma^2_{1,\mathbf{x}}} + \big(\frac{\sigma^2_{1,\mathbf{x}}}{\sigma^2_{2,\mathbf{x}+\mathbf{z}}} - 1\big) + \frac{(\mu_{1,\mathbf{x}} - \mu_{2,\mathbf{x}+\mathbf{z}})^2}{\sigma^2_{2,\mathbf{x}+\mathbf{z}}}\Big) \tag{5.52}$$

Rather than using a parametric form of the intensity densities $p_{I^k,\mathbf{x}}$ and $p_{I^{k-1},\mathbf{x}+\mathbf{z}}$, one can use nonparametric approximations by kernel estimates. For instance:

$$p_{I^k,\mathbf{x}}(u) = \frac{\int_{\mathscr{B}(\mathbf{x},\varepsilon)} \delta(I^k(\mathbf{y}) - u)d\mathbf{y}}{\int_{\mathscr{B}(\mathbf{x},\varepsilon)} d\mathbf{y}} \tag{5.53}$$

and,

$$p_{I^{k-1},\mathbf{x}+\mathbf{z}}(u) = \frac{\int_{\mathscr{B}(\mathbf{x}+\mathbf{z},\varepsilon)} \delta(I^{k-1}(\mathbf{y}) - u)d\mathbf{y}}{\int_{\mathscr{B}(\mathbf{x}+\mathbf{z},\varepsilon)} d\mathbf{y}}, \tag{5.54}$$

where δ is the delta functional. In practice, the image will be discretized to m values $u_j,\ j = 1, ..., m$, and Eq. (5.53) and Eq. (5.54) will give m-bin histograms. One then uses a discrete expression of the Kullback-Leibler divergence (or other density separation function).

Next, we will see how the target shape and motion can be made to intervene to enhance tracking.

5.4.3 Using Shape and Motion

When a model of the target shape is available, it can assist tracking, particularly when occlusion occurs. The inclusion of this model into tracking can be done via an additional term in the objective functional called a *shape prior*. Shape priors in image segmentation have been investigated in several studies [13–15, 68–71] but their application to tracking is scarse [9, 12]. The usefulness of shape priors in tracking was first shown in [9]. The scheme processed multiple targets and allowed targets to occlude each other. A photometric functional was used to detect the targets when there is no occlusion. When an occlusion occurs between targets, an independent external process detects it and a shape functional is consequently instantiated. In [12] a model was learned by viewing the target from different perspectives and determining class-descriptive principal components which were embedded in a shape prior. Tracking was mostly driven by the shape prior in a functional which also contains a photometric data term playing a secondary role. The intensity of the target and its background were modelled by Gaussians.

Functional Eq. (5.40) can be augmented with a shape term that uses the target shape at the previous instant of observation $k-1$ to assist tracking at instant k. As with the photometric appearance, the shape at one instant serves as a model to locate the target at the next instant. This forgoes the need to learn models beforehand as in [12]. Such prior learning may not be possible or practicable because training data is not always available or accessible. Also, the target profile variations due to motion and occlusion may be too complex to circumscribe in a single model. Using the target shape at the previous instant as a model decomposes a possibly significant cumulative target shape variation into a sequence of in-between instants deformations generally simpler to process, affording a better resilience to occlusion and arbitrary shape deformation.

Let the level set function corresponding to a closed regular plane curve be its *signed distance function* (SDF), taken positive in its interior by convention. Let Φ_0 be the SDF level set function of the boundary of R_0 at time k. The squared displaced SDF difference can evaluate the distance between the boundary of a region R and that of R_0 and, therefore serve as a shape tracking term:

$$d^2(\Phi, \Phi_0) = \int_\Omega \Big(H(\Phi(\mathbf{x})) - H(\Phi_0(\mathbf{x}+\mathbf{h}))\Big)^2 d\mathbf{x}, \tag{5.55}$$

where Φ is the SDF level set of R, H is the Heaviside step function, and $\mathbf{h}$ is the motion field between $k-1$ and k. This field can be taken to be an unknown variable in the objective functional. For the purpose of tracking, however, it is sufficient to approximate $\mathbf{h}$ by the average SDF difference because minimizing the functional with respect to the level set, using both the photometric and the shape tracking terms, will iteratively seek adjustments to bring the evolving SDF to coincide with the SDF of R_0 which serves as the current prior. With Eq. (5.55) as a shape tracking term, the objective functional Eq. (5.40) becomes, after rewriting it in terms of Φ to make it consistent with the writing of Eq. (5.55):

$$\begin{aligned} \mathscr{E}(\Phi | I^{k-1}, I^k, \Phi_0) = & \int_\Omega H(\Phi)\ \xi_1(\mathbf{x}) d\mathbf{x} \\ & + \int_\Omega (1 - H(\Phi))\ \xi_2(\mathbf{x}) d\mathbf{x} \\ & + \lambda \int_\Omega |\nabla H(\Phi)| d\mathbf{x} \\ & + \frac{\beta}{2} \int_\Omega (H(\Phi(\mathbf{x})) - H(\Phi_0(\mathbf{x}+\mathbf{h})))^2 d\mathbf{x}, \end{aligned} \tag{5.56}$$

where λ, β are positive weighing constants. The third term on the righthand side above is the usual length regularization term. The corresponding level set evolution equation is:

$$\frac{\partial \Phi}{\partial \tau}(\mathbf{x}, \tau) = -\,\delta(\Phi(\mathbf{x}, \tau))\big[\xi_1(\mathbf{x}) - \xi_2(\mathbf{x}) + \lambda\, \nabla \cdot \frac{\nabla \Phi}{\|\nabla \Phi\|}(\mathbf{x}, \tau) \\ + \beta\, \big(H(\Phi(\mathbf{x}, \tau)) - H(\Phi_0(\mathbf{x} + \mathbf{h}, \tau))\big)\big], \tag{5.57}$$

The equation applies on γ but can be extended to apply on Ω by replacing δ by $\|\nabla \Phi\|$ [72], giving the expression of Eq. (5.36). Alternatively, one can use a regularized approximation H_ε of H [73], which would give:

$$\frac{\partial \Phi}{\partial \tau}(\mathbf{x}, \tau) = -\,\delta_\varepsilon(\Phi(\mathbf{x}, \tau))\big[\xi_1(\mathbf{x}) - \xi_2(\mathbf{x}) + \lambda\, \nabla \cdot \frac{\nabla \Phi}{\|\nabla \Phi\|}(\mathbf{x}, \tau) \\ + \beta\, \big(H_\varepsilon(\Phi(\mathbf{x}, \tau)) - H_\varepsilon(\Phi_0(\mathbf{x} + \mathbf{h}, \tau))\big)\big], \tag{5.58}$$

where $\delta_\varepsilon = H'_\varepsilon$ and τ is the algorithmic time. For instance, one can use the following bounded support approximation [73] for a narrow-band implementation [16]:

$$H_\varepsilon(z) = \begin{cases} 1, & z > \varepsilon \\ 0, & z < \varepsilon \\ \frac{1}{2}\left(1 + \frac{z}{\varepsilon} + \frac{1}{\pi}\sin(\frac{\pi z}{\varepsilon})\right), & |z| \leq \varepsilon \end{cases} \tag{5.59}$$

Algorithm behavior during occlusion: In essence, the shape tracking term complements the photometric tracking term by constraining the active contour at any instant to have a shape as close as possible to the shape of the curve at the preceding instant. This is particularly relevant during occlusion. The algorithm behaves as follows when an occlusion occurs. Suppose the object is visible at some instant and is partially occluded at the next. In the occluded part of the target, the competition for points between the target and the background happens as follows: for a point in this part, the term $\xi_1 - \xi_2$ will most likely be negative because the intensity will most likely be closer to that of a background point than of a target point. $H(\Phi) - H(\Phi_0)$ will be equal to 1 in this case and the shape term will evaluate to β. Therefore, for β sufficiently large, the velocity will be positive and the point will be assigned to the target as it should. When an occlusion is properly resolved, the occluding background segment becomes "part" of the target. It will "move out" of the target only when the occluded part of the target reappears. The behaviour of the algorithm is similar when the occluded part of the target is progressively uncovered. A good value of β to have the algorithm behave in this desired way can be determined experimentally and, in general, remains valid within a given application. In spite of its occlusion accommodation capability, the algorithm has no means to recover the target portions lost to tracking, i.e., which are visible but not recorded as part of the target by the tracking process. Also, as a general purpose tracker, the scheme can be defeated by a sudden onset of significant occlusion.

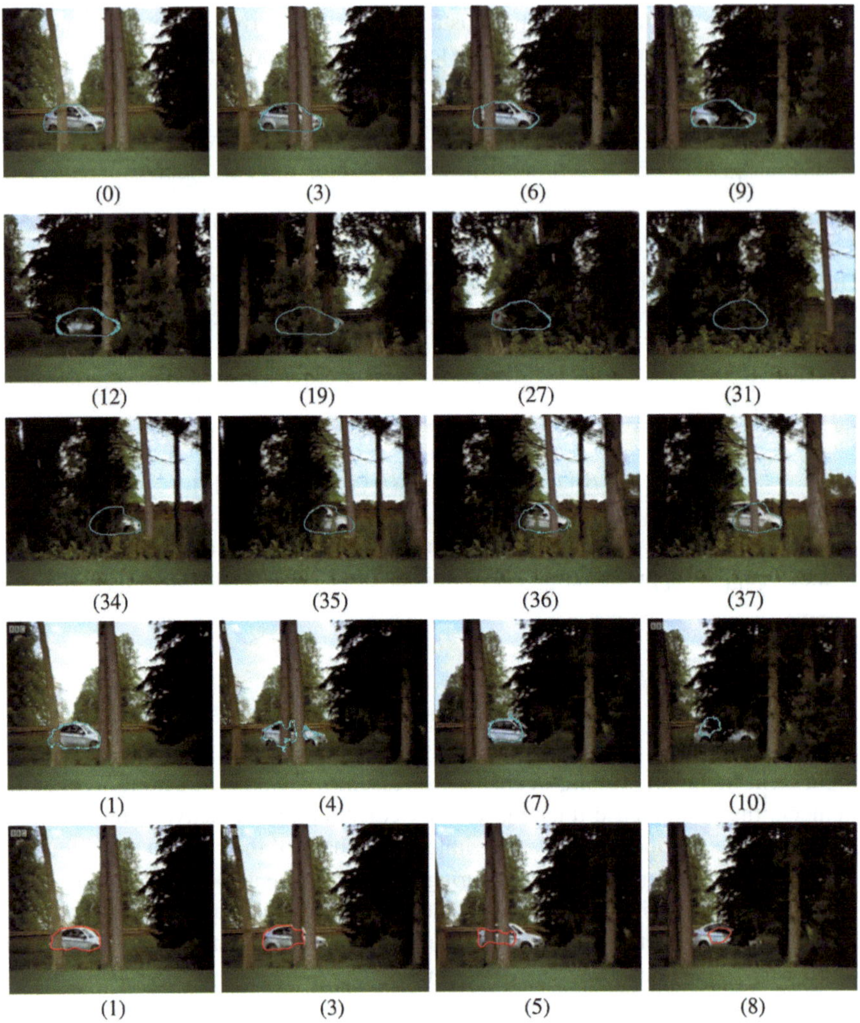

Fig. 5.4 *Car* sequence: The car moves at about constant speed. Partial and total occlusion occur in the sequence. *Top three rows*: Tracking using a shape tracking term. During total occlusion, contour evolution is mainly driven by the shape tracking term. After total occlusion (*third row*) the curve is still on the target although the detected silhouette is inaccurate. *Fourth row*: Tracking with the basic method [52], which does not use shape. *Last row*: Tracking by the tracking distribution method [50] which also does not use shape

Example: The following example [58] illustrates the algorithm during occlusion. It uses the *Car* sequence, where a car has been filmed moving at about constant speed on a road behind a row of trees. There is repeated partial occlusion as well as a moment of total occlusion. The results of tracking with a shape term (Sect. 5.4.3) are depicted in the top three rows of Fig. 5.4. The outcome when not using the shape term [53] (Sect. 5.4.1) is shown in the fourth row. The results with the scheme that tracks model densities of targets [50] (Sect. 5.3), and which also uses no shape information or motion, are in the last row. These results show that shape information and motion were able to assist in keeping track of the target while occlusion occurred.

There are current applications, such as human activity understanding in image sequences, which benefit from processing spatiotemporal information [74, 75]. Along the same vein, one can view tracking as a *spatiotemporal process* which detects the surface generated by the target occluding contour in the $x - y - t$ spatiotemporal domain, rather than a temporally repetitive process, i.e., looking at the problem from the standard viewpoint of locating a target successively from one frame of a sequence to the next. The problem then centers upon a good characterization of the moving target boundaries, which are no other than motion boundaries. In the next section, we will describe a tracking method which generalizes the motion residual subtraction scheme of Sect. 4.6.1, Chap. 4, to the spatiotemporal domain and moving viewing system and objects.

5.5 Tracking in the Spatiotemporal Domain

When objects move in an image sequence, their boundaries carve surfaces in the spatiotemporal domain. The points within the moving objects image regions are animated by motion and thus form the *spatiotemporal motion foreground*. Therefore, tracking can be viewed as motion detection in the spatiotemporal domain [76–78]. Rather than following the target from one instant of observation to the next as we have described so far, the observations in a time interval are all taken together and the spatiotemporal region corresponding to the foreground of moving objects is detected. The moving objects outline can be recovered at any instant t by intersecting this foreground by the plane $\pi(x, y, t) = t$.

The MAP image differencing method in Sect. 4.4.2 of Chap. 4 and its extension to the case of a moving viewing system in Sect. 4.6.1 of the same chapter can be formulated in the spatiotemporal domain [76, 77] by evolving an active surface so that it is brought to coincide with the foreground of moving objects boundary. Here following is the spatiotemporal formulation of the motion residual subtraction method of Sect. 4.6.1, Chap. 4, for tracking in the presence of a moving viewing system. The original spatial formulation and its spatiotemporal extension are obtained in the same manner, have the same structure, and similar equations. The differences are that the spatiotemporal scheme evolves a closed regular surface in 3D space rather than a closed regular plane curve, and the generic derivation of the Euler-Lagrange equations, which deal with surface and volume integrals of scalar functions are more elaborate than for those dealing with the curve and region integrals we have been concerned with so far. These derivations have been described in Chap. 2.

Let $I : (x, y, t) \in D = \Omega \times]0, T[\rightarrow I(x, y, t) \in \mathbb{R}^+$ be an image sequence with open domain $\Omega \subset \mathbb{R}^2$ and time interval of observation $]0, T[$. Let S be a closed regular surface in D to symbolize the boundary of the foreground of moving objects, R_S its interior and $\mathscr{R}_S = \{R_S, R_S^c\}$ the corresponding partition of D. As usual, the complement of the foreground will be called the background. We will allow the viewing system to move and assume that this induces a background motion which can be fully characterized by a vector of parameters $\boldsymbol{\theta}$. Let m be a motion feature. The MAP estimate $(\tilde{S}, \tilde{\boldsymbol{\theta}})$ of $(S, \boldsymbol{\theta})$ using feature m is:

$$\begin{aligned}(\tilde{S}, \tilde{\boldsymbol{\theta}}) &= \arg\max_{S,\boldsymbol{\theta}} P(\mathscr{R}_S, \boldsymbol{\theta}|m) \\ &= \arg\max_{S,\boldsymbol{\theta}} \frac{P(m|\mathscr{R}_S, \boldsymbol{\theta})P(\mathscr{R}_S, \boldsymbol{\theta})}{P(m)}.\end{aligned} \tag{5.60}$$

$P(m)$ is independent of $\boldsymbol{\theta}$ and S and can be ignored. $P(m|\mathscr{R}_S, \boldsymbol{\theta})$ is the data term, and $P(\mathscr{R}_S, \boldsymbol{\theta})$ the *prior*. Assuming conditional independence of the motion feature for $\mathbf{x} \neq \mathbf{y}$, we have:

$$P(m|(\mathscr{R}_S, \boldsymbol{\theta})) = \prod_{\mathbf{x}\in R_S} P(m(\mathbf{x})|\mathscr{R}_S, \boldsymbol{\theta}) \prod_{y\in R_S^c} P(m(\mathbf{y})|\mathscr{R}_S, \boldsymbol{\theta}) \tag{5.61}$$

Maximizing this probability is equivalent to minimizing the negative of its log, which comes to minimizing the functional:

$$\begin{aligned}\mathscr{E}(S, \boldsymbol{\theta}) = &- \int_{R_S} \log P\,(m(\mathbf{x})|(\mathscr{R}_S, \boldsymbol{\theta}))\,d\mathbf{x} \\ &- \int_{R_S^c} log P\,(m(\mathbf{x})|(\mathscr{R}_S, \boldsymbol{\theta}))\,d\mathbf{x} \\ &- \log P(\mathscr{R}_S, \boldsymbol{\theta})\end{aligned} \tag{5.62}$$

The first two terms on the right of Eq. (5.62) can be specified by a data model and the last by a prior.

Data model: Let the *residual normal motion* be the normal component of optical flow from which the normal component of the flow due to viewing system movement has been subtracted:

$$W_*^{\perp} = W^{\perp} - W_c^{\perp}, \tag{5.63}$$

The following data model will constrain the residual to be high inside the spatiotemporal surface generated by the moving objects and low outside, i.e, it will favor high motion activity in the foreground of moving objects and low in the background.

$$P(m(\boldsymbol{x})|\mathscr{R}_S, \boldsymbol{\theta}) \propto \begin{cases} e^{-\alpha e^{-(W_*^{\perp}(\boldsymbol{\theta}))^2}} & \text{for } \boldsymbol{x} \in R_S \\ e^{-\beta(W_*^{\perp}(\boldsymbol{\theta}))^2} & \text{for } \boldsymbol{x} \in R_S^c, \end{cases} \tag{5.64}$$

where $\propto$ is the proportional-to symbol and α and β are positive real constants.

Prior: To bias the foreground surfaces to be smooth, an area-related prior can be used, independent of $\boldsymbol{\theta}$:

$$P(\mathscr{R}_S, \boldsymbol{\theta})) \propto e^{-\lambda \int_S d\sigma} \tag{5.65}$$

These models give the following energy functional to minimize:

$$\mathscr{E}(S, \boldsymbol{\theta}) = \alpha \int_{R_S} e^{-(W_*^{\perp}(\boldsymbol{\theta}))^2} d\rho + \beta \int_{R_S^c} (W_*^{\perp}(\boldsymbol{\theta}))^2 d\rho + \lambda \int_S d\sigma \tag{5.66}$$

This functional involves surface and volume integrals. The Euler-Lagrange equations corresponding to such integrals was derived in Chap. 2.

Describing the image motion induced by the viewing system movement as a translation (a, b), we have, from Eq. (5.63),

$$W_*^{\perp} = -\frac{I_x a + I_y b + I_t}{\|\nabla I\|}$$

Using this expression, the Euler-Lagrange descent equations to minimize objective functional Eq. (5.66) are given by:

$$\begin{cases} \frac{\partial a}{\partial \tau} = \alpha \int_{R_S} \frac{2I_x}{\|\nabla I\|} (\frac{I_t + aI_x + bI_y}{\|\nabla I\|}) e^{-(\frac{I_t + aI_x + bI_y}{\|\nabla I\|})^2} d\rho \\ \quad -\beta \int_{R_S^c} \frac{2I_x}{\|\nabla I\|} (\frac{I_t + aI_x + bI_y}{\|\nabla I\|}) d\rho \\ \frac{\partial b}{\partial \tau} = \alpha \int_{R_S} \frac{2I_y}{\|\nabla I\|} (\frac{I_t + aI_x + bI_y}{\|\nabla I\|}) e^{-(\frac{I_t + aI_x + bI_y}{\|\nabla I\|})^2} d\rho \\ \quad -\beta \int_{R_S^c} \frac{2I_y}{\|\nabla I\|} (\frac{I_t + aI_x + bI_y}{\|\nabla I\|}) d\rho \\ \frac{\partial S}{\delta \tau} = -(2\lambda\kappa + \alpha e^{-(W_*^{\perp}(a,b))^2} - \beta (W_*^{\perp}(a, b))^2)\mathbf{n}, \end{cases} \tag{5.67}$$

where $\mathbf{n}$ is the unit normal function of S and κ its curvature function. The descent equation for the surface, the last equation in Eq. (5.67), was derived using the calculus for surface and volume integrals in Chap. 2 and assuming that the motion parameters a and b are independent of the active surface S. The corresponding level set equation has the usual form except that the level set function is a function of both space and time and, therefore, S is represented implicitly as the zero level of a one-parameter family of functions ϕ, indexed by algorithmic time τ:

$$(\forall \tau) \quad \phi(x(\tau), y(\tau), t(\tau), \tau) = 0, \tag{5.68}$$

where x, y, and t are the spatiotemporal coordinates, and the level set evolution equation is given by:

$$\frac{\partial \phi}{\partial \tau} = -(2\lambda\kappa + \alpha e^{-(W_*^{\perp}(a,b))^2} - \beta (W_*^{\perp}(a, b))^2)) \|\nabla \phi\| \tag{5.69}$$

The initial position S_0 of active surface S is chosen so as to subsume the volume generated by the moving objects. Some practical rules on how to chose the coefficients α, β, and λ are given in [77]. With the proper choice of these, the surface evolves as follows. In the background, we will have $W_*^{\perp} \approx 0$ and the speed V of the movement

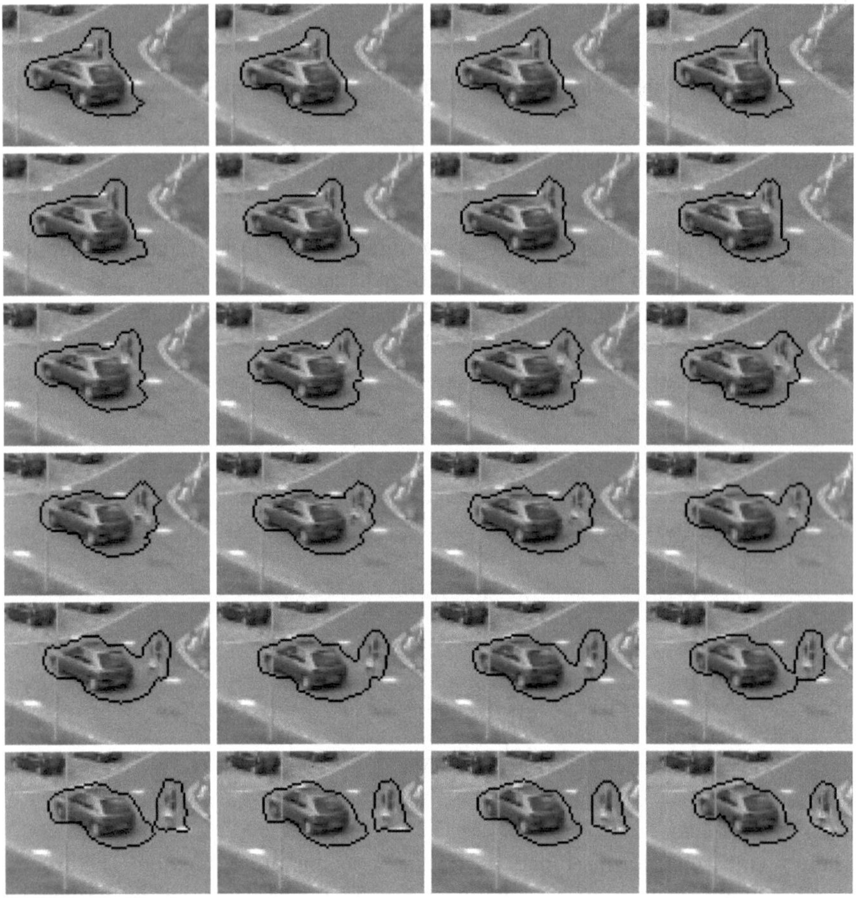

Fig. 5.5 Tracking through the *Car-and-pedestrian* sequence by spatiotemporal motion detection, one frame interval from upper-left to lower-right. The car depicted in this figure drives by a walking pedestrian. The camera moves and induces an approximately translational motion of the image from one frame to the next. The motion boundary englobes initially both the car and the pedestrian it occludes; when the occlusion ends, this boundary splits into two distinct boundaries, one each for the car and pedestrian

of S the direction of its normal will be approximately $V \approx (-2\lambda\kappa - \alpha)\mathbf{n}$ and, therefore, S will move inward. When it reaches the foreground boundary, we will have $V \approx (-2\lambda\kappa + \beta(W_*^{\perp}(a, b))^2)\mathbf{n}$. The term $\beta|W_*^{\perp}|$ acts to prevent the surface from penetrating into the foreground where motion occurs. The term $-2\lambda\kappa$ has a spatiotemporal smoothing effect on the active surface S.

Example: The following example [79] uses the *Car-and-pedestrian* sequence of a car driving by a walking pedestrian. The camera moves to cause an approximately translational image motion. Figure 5.5 presents the results of tracking. Note how

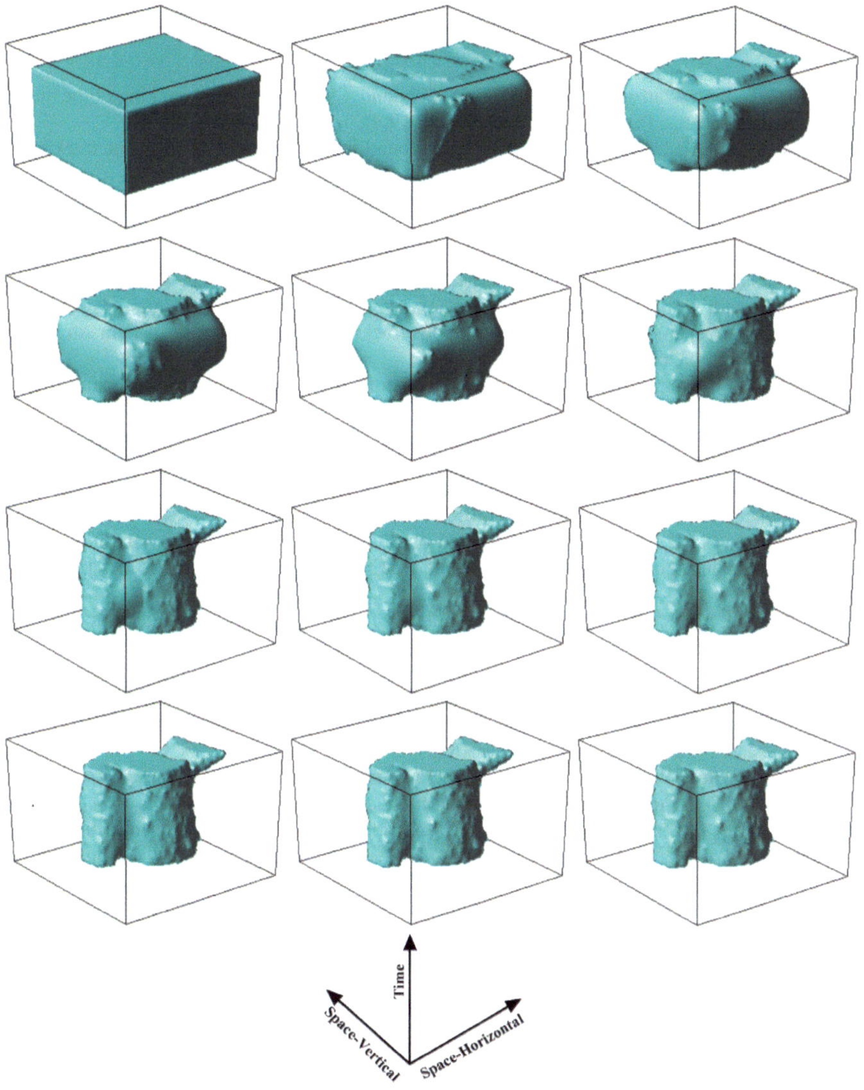

Fig. 5.6 The surface spatiotemporal evolution for the *Car-and-pedestrian* sequence. A cut at time t of the spatiotemporal surface perpendicular to the time axis gives the foreground of the moving objects at time t

the motion boundary which initially encompasses both the pedestrian and the car which occludes it, later splits into two distinct boundaries, one each for the car and pedestrian, when the occlusion ends. The spatiotemporal surface evolution is illustrated in Fig. 5.6.

References

1. C.R. Wren, A. Azarbayejani, T. Darrell, A. Pentland, Pfinder: real-time tracking of the human body. IEEE Trans. Pattern Anal. Mach. Intell. **19**(7), 780–785 (1997)
2. C. Stauffer, W.E.L. Grimson, Adaptive background mixture models for real-time tracking, in *CVPR*, 1999, pp. 2246–2252
3. A. Elgammal, R. Duraiswami, D. Harwood, L. Davis, Background and foreground modeling using nonparametric kernel density estimation for visual surveillance. Proc. IEEE **90**, 1151–1163 (2002)
4. D. Comaniciu, V. Ramesh, P. Meer, Kernel-based object tracking. IEEE Trans. Pattern Anal. Mach. Intell. **25**(5), 564–575 (2003)
5. J. Kang, I. Cohen, G.G. Medioni, Continuous tracking within and across camera streams, in *CVPR (1)*, 2003, pp. 267–272
6. J.-M. Geusebroek, A.W.M. Smeulders, A six-stimulus theory for stochastic texture. Int. J. Comput. Vision **62**(1–2), 7–16 (2005)
7. L. Li, M.K.H. Leung, Integrating intensity and texture differences for robust change detection. IEEE Trans. Image Process. **11**(2), 105–112 (2002)
8. A. Monnet, A. Mittal, N. Paragios, V. Ramesh, Background modeling and subtraction of dynamic scenes, in *ICCV*, 2003, pp. 1305–1312
9. A. Yilmaz, X. Li, M. Shah, Contour-based object tracking with occlusion handling in video acquired using mobile cameras. IEEE Trans. Pattern Anal. Mach. Intell. **26**(11), 1531–1536 (2004)
10. S. Birchfield, S. Rangarajan, Spatial histograms for region-based tracking. Electron. Telecommun. Res. (ETRI) J. **29**(5), 697–699 (2007)
11. C. Veenman, M. Reinders, E. Backer, Resolving motion correspondence for densely moving points. IEEE Trans. Pattern Anal. Mach. Intell. **23**(1), 54–72 (January 2001)
12. D. Cremers, Dynamical statistical shape priors for level set-based tracking. IEEE Trans. Pattern Anal. Mach. Intell. **28**(8), 1262–1273 (August 2006)
13. M.E. Leventon, W.E.L. Grimson, O. Faugeras, Statistical shape influence in geodesic active contours. IEEE Conf. Comput. Vision Pattern Recogn. **1**, 316–323 (2000)
14. D. Cremers, S. Soatto, A pseudo distance for shape priors in level set segmentation, in *IEEE International Workshop on Variational Geometric and Level Set Methods*, 2003, pp. 169–176
15. X. Bresson, P. Vandergheynst, J. Thiran, A variational model for object segmentation using boundary information and shape prior driven by the mumford-shah functional. Int. J. Comput. Vision **68**(2), 145–162 (2006)
16. J.A. Sethian, *Level Set Methods and Fast Marching Methods* (Cambridge University Press, Cambridge, 1999)
17. F. Meyer, P. Bouthemy, Region-based tracking using affine motion models in long image sequences. Comput. Vis. Graph. Image Process. **60**(2), 119–140 (1994)
18. Y. Mae, Y. Shirai, J. Miura, Y. Kuno, Object tracking in cluttered background based on optical flow and edges, in *International Conference on Pattern recognition*, vol. I, 1996, pp. 196–200
19. M. Bertalmio, G. Sapiro, G. Randall, Morphing active contours. IEEE Trans. Pattern Anal. Mach. Intell. **22**(7), 733–737 (2000)
20. S. Jehan-Besson, M. Barlaud, G. Aubert, Detection and tracking of moving objects using a new level set based method, in *International Conference on Pattern Recognition*, (Barcelona, Spain, Sept. 2000), pp. 7112–7117
21. H. Tsutsui, J. Miura, Y. Shirai, Optical flow-based person tracking by multiple cameras, in *Multisensor Fusion and Integration for Intelligent Systems, MFI 2001. International Conference on*, 2001, pp. 91–96
22. N. Paragios, R. Deriche, Geodesic active contours and level sets for the detection and tracking of moving objects. IEEE Trans. Pattern Anal. Mach. Intell. **22**(3), 266–280 (2000)
23. T. Brox, B. Rosenhahn, D. Cremers, H.-P. Seidel, High accuracy optical flow serves 3-D pose tracking: Exploiting contour and flow based constraints, in *European Conference on Computer Vision*, 2006, pp. 98–111

24. C. Harris, M. Stephens, A combined corner and edge detector, in *Alvey Vision Conference*, 1988, pp. 147–152
25. C. Tomasi, J. Shi, Good features to track, in *Computer Vision and Pattern Recognition Conference*, 1994, pp. 593–600
26. D.G. Lowe, Distinctive image features from scale-invariant keypoints. Int. J. Comput. Vision **60**(2), 91–110 (2004)
27. J.K. Aggarwal, R.O. Duda, Computer analysis of moving polygonal images. IEEE Trans. Comput. **c-24**(10), 966–976 (1975)
28. W.K. Chow, J.K. Aggarwal, Computer analysis of planar curvilinear moving images. IEEE Trans. Comput. **c-26**(2), 179–185 (1977)
29. I. Sethi, R. Jain, Finding trajectories of feature points in a monocular image sequence. IEEE Trans. Pattern Anal. Mach. Intell. **9**(1), 56–73 (1987)
30. J. Crowley, P. Stelmaszyk, T. Skordas, P. Puget, C. Discours, Measuring image flow by tracking edge-lines, in *IEEE International Conference on Computer Vision*, 1988, pp. 658–664
31. O. Faugeras, R. Deriche, Tracking line segments, in *European Conference on Computer Vision*, 1990, pp. 259–268
32. Y. Bar-Shalom, *Tracking and data association* (Academic Press Professional, Inc., San Diego, CA, USA, 1987)
33. I.J. Cox, A review of statistical data association techniques for motion correspondence. Int. J. Comput. Vision **10**(1), 53–66 (1993)
34. T. Broida, R. Chellappa, Estimation of object motion parameters from noisy images. IEEE Trans. Pattern Anal. Mach. Intell. **8**(1), 90–99 (January 1986)
35. R. Deriche, O. Faugeras, Tracking line segments. Image Vis. Comput. **8**(4), 261–270 (1990)
36. A. Blake, R. Curwen, A. Zisserman, A framework for spatiotemporal control in the tracking of visual contours. Int. J. Comput. Vision **11**, 127–145 (1993). 10.1007/BF01469225. http://dx.doi.org/10.1007/BF01469225
37. B. Bascle, P. Bouthemy, R. Deriche, F. Meyer, Tracking complex primitives in an image sequence, in *International Conference on Pattern Recognition*, vol. 1, 1994, pp. 426–431
38. C. Kervrann, F. Heitz, Robust tracking of stochastic deformable models in long image sequences, in *International Conference on Image Processing*, 1994, pp. 88–92
39. N. Peterfreund, Robust tracking with spatio-velocity snakes: Kalman filtering approach, in *ICCV*, 1998, pp. 433–439
40. R. Rosales, S. Sclaroff, 3D trajectory recovery for tracking multiple objects and trajectory guided recognition of actions, in *Computer Vision and Pattern Recognition Conference*, 1999, pp. 2117–2123
41. Y. Boykov, D.P. Huttenlocher, Adaptive bayesian recognition in tracking rigid objects, in *Computer Vision and Pattern Recognition Conference*, 2000, pp. 697–704
42. N. Gordon, D. Salmond, A. Smith, A novel approach to non-linear and non-gaussian bayesian state estimation. IEE Proc. F **140**, 107–113 (1993)
43. S. Arulampalam, S. Maskell, N. Gordon, T. Clapp, A tutorial on particle filters for on-line non-linear/non-gaussian bayesian tracking. IEEE Trans. Signal Process. **50**(2), 174–188 (2002)
44. M. Isard, A. Blake, Condensation-conditional density propagation for visual tracking. Int. J. Comput. Vision **29**(1), 5–28 (1998)
45. U. Grenander, Y. Chow, D. Keenan, *Hands: A Pattern Theoretic Study Of Biological Shapes* (Springer, New York, 1991)
46. E. Anderson, *Monte Carlo Methods and Importance Sampling, Lecture Notes for Stat 378C* (University of California, Berkeley, 1999), pp. 1–8
47. B. Li, R. Chellappa, Simultaneous tracking and verification via sequential posterior estimation, in *IEEE Conference on Computer Vision and Pattern Recognition*, 2000, pp. 110–117
48. C. Hue, J. Cadre, P. Perez, Sequential monte carlo filtering for multiple target tracking and data fusion. IEEE Trans. Signal Process. **50**(2), 309–325 (2002)
49. D. Comaniciu, P. Meer, Mean shift: a robust approach toward feature space analysis. IEEE Trans. Pattern Anal. Mach. Intell. **24**(5), 603–619 (2002)

50. D. Freedman, T. Zhang, Active contours for tracking distributions. IEEE Trans. Image Process. **13**(4), 518–526 (2004)
51. A. Mitiche, I. Ben Ayed, *Variational and Level Set Methods in Image Segmentation* (Springer, New York, 2010)
52. A. Mansouri, Region tracking via level set pdes without motion computation. IEEE Trans. Pattern Anal. Mach. Intell. **24**(7), 947–961 (2002)
53. A. Mansouri, A. Mitiche, Region tracking via local statistics and level set pdes, in *Conference on Image Processing*, vol. III, ed. by I.E.E.E. International (USA, Rochester, NY, 2002), pp. 605–608
54. V. Caselles, R. Kimmel, G. Sapiro, Geodesic active contours. Int. J. Comput. Vision **22**(1), 61–79 (1997)
55. S. Jehan-Besson, M. Barlaud, G. Aubert, Detection and tracking of moving objects using a new level set based method, in *ICPR*, 2000, pp. 7112–7117
56. R.J. Radke, S. Andra, O. Al-Kofahi, B. Roysam, Image change detection algorithms: a systematic survey. IEEE Trans. Image Process. **14**(3), 294–307 (2005)
57. T. Bouwmans, F.E. Baf, B. Vachon, Background modelling using mixture of gaussians for foreground detection. IEEE Trans. Image ProceRecent Pattents Comput. Sci. **1**(3), 219–237 (2008)
58. M. Ben Salah, Fonctions noyaux et a priori de forme pour la segmentation d'images et le suivi d'objets, Ph.D. dissertation, Institut national de la recherche scientifique, INRS-EMT, 2011
59. M. Ben Salah, A. Mitiche, Model-free, occlusion accommodating active contour tracking. ISRN Artif. Intell. **2012**, Article ID 672084, 15 (2012)
60. R.O. Duda, P.E. Hart, *Pattern Classification and Scene Analysis* (John Wiley & Sons, New York, 1973)
61. R.O. Duda, P.E. Hart, D.G. Stork, *Pattern Classification* (Wiley, New York, 2000)
62. B.U. Park, J.S. Marron, Comparison of data-driven bandwidth selectors. J. Am. Stat. Assoc. **85**, 66–72 (1990)
63. K. Fukunaga, L. Hostetler, The estimation of the gradient of a density function, with applications in pattern recognition. IEEE Trans. Inf. Theory **21**(1), 32–40 (1975)
64. D. Comaniciu, P. Meer, Mean shift analysis and applications. in *International Conference on Computer Vision*, 1999, pp. 1197–1203
65. D. Comaniciu, V. Ramesh, P. Meer, Real-time tracking of non-rigid objects using mean shift, in *Computer Vision and Pattern Recognition Conference*, 2000, pp. 2142–2149
66. P. Martin, P. Refregier, F. Goudail, F. Guerault, Influence of the noise model on level set active contour segmentation. IEEE Trans. Pattern Anal. Mach. Intell. **26**(6), 799–803 (2004)
67. T. Zhang, D. Freedman, Improving performance of distribution tracking through background mismatch. IEEE Trans. Pattern Anal. Mach. Intell. **27**(2), 282–287 (2005)
68. Y. Chen, H. Tagare, S.R. Thiruvenkadam, F. Huang, D.C. Wilson, K.S. Gopinath, R.W. Briggs, E.A. Geiser, Using prior shapes in geometric active contours in a variational framework. Int. J. Comput. Vision **50**(3), 315–328 (2002)
69. M. Rousson, N. Paragios, Shape priors for level set representations. Eur. Conf Comput. Vision, 2002, pp. 416–418
70. A. Tsai, A.J. Yezzi, W.M. Wells III, C.M. Tempany, D. Tucker, A.C. Fan, W.E.L. Grimson, A.S. Willsky, A shape-based approach to the segmentation of medical imagery using level sets. IEEE Trans. Med. Imaging **22**(2), 137–154 (2003)
71. T. F. Chan, W. Zhu, Level set based shape prior segmentation, in *IEEE Conference on Computer Vision and, Pattern Recognition*, 2005, pp. 1164–1170
72. H.-K. Zhao, T. Chan, B. Merriman, S. Osher, A variational level set approach to multiphase motion. J. Comput. Phys. **127**(1), 179–195 (1996)
73. T. Chan, L. Vese, Active contours without edges. IEEE Trans. Image Process. **10**(2), 266–277 (2001)
74. M.S. Ryoo, J.K. Aggarwal, Spatio-temporal relationship match: Video structure comparison for recognition of complex human activities, in *International Conference on Computer Vision*, 2009, pp. 1–8

75. M.S. Ryoo, J.K. Aggarwal, Human activities: Handling uncertainties using fuzzy time intervals, in *International Conference on Image Processing*, 2008, pp. 1–4
76. A. Mitiche, R. Feghali, A. Mansouri, Motion tracking as spatio-temporal motion boundary detection. J. Robot. Auton. Systems **43**, 39–50 (2003)
77. R. El-Feghali, A. Mitiche, Spatiotemporal motion boundary detection and motion boundary velocity estimation for tracking moving objects with a moving camera: a level sets pdes approach with concurrent camera motion compensation. IEEE Trans. Image Process. **13**(11), 1473–1490 (2004)
78. M. Ristivojevic, J. Konrad, Space-time image sequence analysis: object tunnels and occlusion volumes. IEEE Trans. Image Process. **15**(2), 364–376 (2006)
79. R. Feghali, Tracking of moving objects in an image sequence via active spatiotemporal surfaces, Ph.D. dissertation, Institut National de la recherche scientifique, INRS-EMT, 2003

Chapter 6
Optical Flow Three-Dimensional Interpretation

6.1 Introduction

Optical flow is the field of optical velocity vectors of the projected environmental surfaces whenever a viewing system moves relative to the viewed environment. Therefore, optical flow carries information about the imaged surfaces and their movement [1–3]. The object of *three-dimensional* (3D) *interpretation* of optical flow is to recover the structure and motion of these surfaces and segment the environment into differently moving objects.

Sparse interpretation is usually distinguished from *dense*. Sparse interpretation recovers the 3D variables, namely depth and motion, for a few points. These correspond to characteristic image points, points that can easily and consistently be identified in distinct views of the environment. Dense methods, by contrast, seek to infer depth and 3D motion for all the points on the visible surfaces. Sparse methods were investigated first (see [4–6] for bibliographies and [7] for reprints of early papers on structure from point correspondences [8, 9]) because the fundamental point-wise projective relationships between the environment and its image had to be discerned before dense interpretation could be addressed.

Interpretation is also often referred to as *direct* or *indirect*. It is indirect when optical flow is estimated and used explicitly as data by the 3D recovery process. Optical flow can be estimated independently of 3D interpretation but can also be done concurrently. Direct recovery originated with [10], followed by [11, 12]. By substituting the variables of a 3D model for optical flow, for instance the Longuet-Higgins and Prazdny rigid-motion model which we will discuss in the next section [13–15], optical velocities will no longer appear explicitly in the recovery process and the interpretation is called direct. The question of whether interpretation is direct or indirect has been asked about the human visual system and there is evidence that environmental motion perception is an indirect process involving two separate steps, retinal motion evaluation and 3D interpretation [16].

Dense interpretation is considerably more complex than sparse, but the Longuet-Higgins and Prazdny model, the work of Horn and Shuck on optical flow

A. Mitiche and J. K. Aggarwal, *Computer Vision Analysis of Image Motion by Variational Methods*, Springer Topics in Signal Processing 10, DOI: 10.1007/978-3-319-00711-3_6,

estimation [17], and recent variational and level set statements of fundamental vision problems [18–21] have opened up the possibility of effective processing.

Several investigations of dense interpretation have addressed the case of a viewing system moving in an otherwise static environment [12, 22–33]. This case simplifies the problem significantly because the single 3D motion to recover is that of the viewing system. Also, segmentation of the environment with respect to motion is not an issue and this simplifies the problem further.

When the environmental objects and the viewing system are allowed to move independently, it is essential that motion boundaries be included in the interpretation so that the moving objects can be delineated accurately. Motion boundary preservation is a central issue in 3D interpretation of optical flow.

The simultaneous movement of the viewing system and the viewed objects in dense interpretation has been addressed in several studies [34–41]. The investigations in [34–36] are non variational methods which assume that optical flow is given before interpretation. They address motion segmentation by grouping processes such as region growing by 3D motion [36], clustering of 3D motion via mixture models [34], and clustering via oriented projections of optical flow [35]. By assuming that optical flow is available before interpretation, the methods put the burden on optical flow estimation to place motion boundaries accurately.

The methods investigated in [37–41] are variational methods. Their functionals all contain a data term which uses the Longuet-Higgins and Prazdny rigid motion model and all have a provision for preserving 3D motion and depth boundaries. As described briefly in the following, they differ in the way these boundaries are described.

The minimum description length (MDL) discrete scheme of [37] is a transcription to optical flow 3D interpretation of the MDL piecewise constant image segmentation of Leclerc [42]. MDL encoding refers to local edges rather than boundaries as curves. This lack of explicit global region boundary information generally leads to fragmented segmentation.

In a continuous formulation, 3D interpretation boundaries can be preserved via a length regularization term which would allow smoothing along the boundaries and inhibit it across, very much like what have done some of the optical flow estimation methods we studied in Chap. 3. In such a framework, the formulation of [38] minimizes an integral over the image domain containing a data term, based on the Longuet-Higgins and Prazdny rigid motion model, and a term of regularization by anisotropic diffusion to preserve depth discontinuities. Motion segmentation is not addressed explicitly.

Along a different vein of thought, motion boundaries can be accounted for by an active curves functional for joint optical flow 3D segmentation and 3D interpretation, in which case the segmentation will refer explicitly to the active curves as the 3D interpretation boundaries. Such an approach was investigated in [39]. The objective functional contained a data term for each segmentation region and terms of regularity of the regions boundary and depth. Minimization of the objective functional led to segmentation by curve evolution and concurrent nonlinear estimation of relative depth. Along the same vein, concurrent optical flow estimation and 3D interpretation has been addressed within the active curve segmentation framework in [40, 41]. Joint estimation allowed the linearized expression of the Longuet-Higgins and Prazdny rigid motion model in the data term, leading to linear 3D motion estimation within the segmentation regions. Segmentation was based on 3D motion in [40] and on optical flow in [41].

Finally, 3D interpretation of optical flow can be done via scene flow estimation, where no model of motion, rigid or other, need to be assumed. Scene flow is the field of 3D velocities of the visible environmental surfaces. Scene flow has been studied mainly in stereoscopy [43–47], although it stands independent of stereoscopy. In (Sect. 6.5) we will describe how it can be recovered independently of stereoscopy by a scheme reminiscent of the Horn-and-Schunck optical flow estimation method.

The purpose of this chapter is to address dense 3D interpretation of optical flow with an emphasis on its central issues, specifically motion boundary preservation and motion segmentation. We will start (Sect. 6.2) by specifying the imaging model we will be using, namely central projection in a Cartesian reference system, for which we will write the 3D-to-2D equations of projection. We will also write the equations which relate optical flow to the variables of 3D rigid structure and motion, particularly the Longuet-Higgins and Prazdny fundamental model. All of these basic equations will be used repeatedly in subsequent discussions. We will follow (Sect. 6.3) with the study of 3D interpretation for a viewing system moving in a static environment. In this case, the movement of the viewing system is the only motion to recover, which simplifies the problem significantly. The case where the environmental objects can also move introduces not only the unknowns of structure and motion of each moving object but also the problem of segmenting the environment into differently moving objects. In this context, point-wise as well as region-based dense variational methods will be described (Sect. 6.4). The chapter will conclude with scene flow estimation (Sect. 6.5).

Before we set out to describe methods, we must to point out that 3D interpretation of optical flow has intrinsic limitations: (a) An obvious limitation is that depth cannot be recovered for surfaces which do not move relative to the viewing system. Also, depth of untextured surfaces cannot be recovered, unless by some form of regularization; (b) to a 3D interpretation consistent with a sequence spatiotemporal variations corresponds a family of scaled interpretations; (c) reference to optical flow implies small-range image motion. For long-range motion, multiresolution/multigrid processing must support the interpretation.

6.2 Projection Models

The viewing system will be modelled by a direct orthonormal reference system $\mathscr{S} = (\mathbf{O}; \mathbf{I}, \mathbf{J}, \mathbf{K})$, where $\mathbf{I}, \mathbf{J}, \mathbf{K}$ are the unit vectors on the X-, Y- and Z-axes, and central projection through the origin $\mathbf{O}$. The Z-axis is the axis of depth. The image plane π is perpendicular to the Z-axis at distance f (the focal length) from the origin.

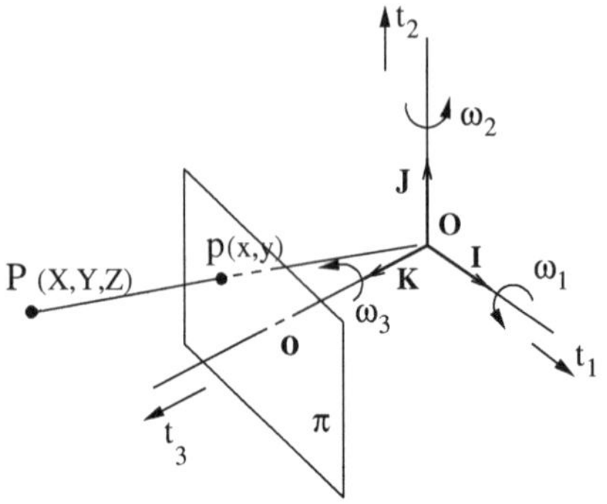

Fig. 6.1 The viewing system is symbolized by an orthonormal direct reference system $\mathscr{S} = (\mathbf{O}; \mathbf{I}, \mathbf{J}, \mathbf{K})$ and central projection through the origin **O** on image plane π parallel to plane $P_{\mathbf{IJ}}$ and at focal distance f from **O**. Point **P** in space has coordinates X, Y, Z. Its projection, or image, has coordinates $x = f\frac{X}{Z}$ and $y = f\frac{Y}{Z}$. If **P** is a point of a rigid body, its velocity $\mathbf{P}'$, i.e., the derivative of its position with respect to time $\frac{d\mathbf{P}}{dt}$, has instantaneous translational and rotational components $\mathbf{T} = (\tau_1, \tau_2, \tau_3)$, $\omega = (\omega_1, \omega_2, \omega_3)$ according to the fundamental formula of rigid body motion: $\mathbf{P}' = \mathbf{T} + \omega \times \mathbf{OP}$

The configuration is drawn in Fig. 6.1. Let **P** be a point on an environmental surface, (X, Y, Z) its coordinates in $\mathscr{S}$, and x, y, f the coordinates of its projection **p** on π.

Points **P**, **p**, and **O** are aligned. Therefore:

$$\frac{X-0}{x-0} = \frac{Y-0}{y-0} = \frac{Z-0}{f-0}, \tag{6.1}$$

which gives the coordinate projection equations:

$$\begin{aligned} x &= f\frac{X}{Z} \\ y &= f\frac{Y}{Z}. \end{aligned} \tag{6.2}$$

When **P** moves relative to the viewing system, its coordinates and those of its projection **p** are functions of time. Differentiation of both sides of the coordinate projection equations Eq. (6.2) gives the equations connecting optical flow to 3D motion and depth:

$$\begin{aligned} u &= f\frac{U - xW}{Z} \\ v &= f\frac{V - yW}{Z}, \end{aligned} \tag{6.3}$$

where $U = \frac{dX}{dt}$, $V = \frac{dY}{dt}$, $W = \frac{dZ}{dt}$ are the coordinates of the velocity of **P**, i.e., of 3D motion, and u, v are those of the velocity of the projection **p** of **P**, i.e., of optical flow at **p**. When the surface on which **P** lies is a rigid body, the 3D motion takes the particular form of a kinematic screw in space, i.e., the composition of a translation and a rotation about an axis through an arbitrary point. When this point is the reference system origin, **O**, the 3D velocity separates into the constituent instantaneous translational and rotational parts, **T** and ω, respectively, according to the fundamental formula:

$$\frac{d\mathbf{P}}{dt} = \mathbf{T} + \omega \times \mathbf{OP}, \tag{6.4}$$

which expands as:

$$\begin{aligned} U &= \tau_1 + Z\omega_2 - Y\omega_3 \\ V &= \tau_2 + X\omega_3 - Z\omega_1 \\ W &= \tau_3 + Y\omega_1 - X\omega_2, \end{aligned} \tag{6.5}$$

where τ_1, τ_2, τ_3 are the components of **T** and $\omega_1, \omega_2, \omega_3$ are those of ω. If we substitute the coordinates projection relations Eq. (6.2) in Eq. (6.5) and the resulting expressions in Eq. (6.3), we obtain the Longuet-Higgins and Prazdny model equations which connect optical flow to depth and rigid body motion in space:

$$\begin{aligned} u &= -\tfrac{xy}{f}\omega_1 + \tfrac{f^2+x^2}{f}\omega_2 - y\omega_3 + \tfrac{f\tau_1 - x\tau_3}{Z} \\ v &= -\tfrac{f^2+y^2}{f}\omega_1 + \tfrac{xy}{f}\omega_2 + x\omega_3 + \tfrac{f\tau_2 - y\tau_3}{Z}. \end{aligned} \tag{6.6}$$

If rigid motion $(\mathbf{T}, \omega)$ and scalar field Z verify these rigid motion equations then $\alpha\mathbf{T}, \omega, \alpha Z,\ \alpha \in \mathbb{R}^+$, also verify them. This uncertainty of scale in 3D rigid body motion interpretation is manifest in the more general expressions Eq. (6.3) which are verified only up to a multiplication of the 3D velocity and depth by a (same) scale factor. Sparse methods of optical flow 3D interpretation which use the Longuet-Higgins and Prazdny model equations generally "fix" this scale by fixing the norm of the rigid body translational component, $\|\mathbf{T}\|$, say equal to 1, which means that only the direction of translation can be recovered. Scale can also be fixed by arbitrarily fixing the depth of one of the rigid body points.

A count of unknowns and equations in Eq. (6.6) shows that sparse methods require the observation of at least five points on the same rigid body and their optical flow values. Five points would give 10 equations, and 5 unknowns of depth, 6 unknowns for the screw of motion, minus 1 unknown for fixing scale, for a total of 10 unknowns. Any distinct additional point on the rigid body adds two equations and one unknown of depth.

The rigid body motion equations Eq. (6.6), which are nonlinear in Z, can be linearized by eliminating depth. This can be done by pulling depth out of each of these equations, equating the resulting expressions to get an equation in the components of translation and rotation, and making the following change of variables:

$$e_1 = -(\omega_3\tau_3 + \omega_2\tau_2)$$

$$\begin{aligned}
e_2 &= -(\omega_3 \tau_3 + \omega_1 \tau_1) \\
e_3 &= -(\omega_2 \tau_2 + \omega_1 \tau_1) \\
e_4 &= \omega_2 \tau_1 + \omega_1 \tau_2 \\
e_5 &= \omega_3 \tau_1 + \omega_1 \tau_3 \\
e_6 &= \omega_3 \tau_2 + \omega_2 \tau_3,
\end{aligned} \tag{6.7}$$

and:

$$\begin{aligned}
e_7 &= \tau_1 \\
e_8 &= \tau_2 \\
e_9 &= \tau_3,
\end{aligned} \tag{6.8}$$

leading to the homogeneous linear equation:

$$< \mathbf{d}, \mathbf{e} > = 0, \tag{6.9}$$

where $\mathbf{e} = (e_1, e_2, \ldots, e_9)^T$ and $\mathbf{d}$ is the data vector, which is a function of image position and optical flow:

$$\mathbf{d} = (x^2, y^2, f^2, xy, fx, fy, -fv, fu, -yu + xv)^T. \tag{6.10}$$

Vector $\mathbf{e}$ characterizes the 3D motion of a rigid body, i.e., it applies to all the points of the same rigid body. It is called the vector of *essential parameters* of the rigid body. The homogeneity of Eq. (6.9) reflects the uncertainty of scale we mentioned previously. A count of unknowns and equations, accounting for scale, shows that sparse methods of 3D interpretation which use Eq. (6.9) require the observation of at least 8 points on the same rigid body and their optical flow values. Recovery of the essential parameter vector $\mathbf{e}$ uniquely determines the original variables $\mathbf{T}$, ω.

The relations between 3D interpretation and image variables we have so far written, namely Eq. (6.3); the Longuet-Higgins and Prazdny model Eq. (6.6); and the linearized version Eq. (6.9); all involve image position and optical flow but not the image sequence itself explicitly. However, optical flow is not a sensed variable and must be estimated using the image sequence, for instance by one of the methods described in Chap. 3. To obtain a 3D interpretation equation in which the image sequence function appears, via its spatiotemporal derivatives, but not optical flow, one can substitute for optical flow in the Horn and Schunck equation [17]:

$$I_x u + I_y v + I_t = 0, \tag{6.11}$$

its expression in terms of the 3D variables of an interpretation model, for instance one of the model equations we have just mentioned. For example, if we write u and v according to the Longuet-Higgins and Prazdny model, we obtain the Negahdaripour 3D rigid motion constraint [11]:

$$\frac{1}{Z} < \mathbf{s}, \mathbf{T} > + < \mathbf{q}, \omega > + I_t = 0, \tag{6.12}$$

where vectors **s**, and **q** are given by

$$\mathbf{s} = \begin{bmatrix} f I_x \\ f I_y \\ -x I_x - y I_y \end{bmatrix}, \quad \mathbf{q} = \begin{bmatrix} -f I_y - \frac{y}{f}(x I_x + y I_y) \\ f I_x + \frac{x}{f}(x I_x + y I_y) \\ -y I_x + x I_y \end{bmatrix} \tag{6.13}$$

In the upcoming sections we will see how 3D interpretation model equations can be used in a variational formulation of dense 3D interpretation of optical flow. We will begin with ego-motion in a static environment, the simplest case, and then treat relative motion of viewing system and viewed objects.

6.3 Ego-Motion in a Static Environment

Ego-motion is the movement of the observer, in our case the viewing system, or camera. Ego-motion in a static environment presents 3D interpretation with the simplest problem because the viewing system rigid movement is the only motion to recover. In spite of this simple presentation, the subject has received a considerable attention. The depth scalar field is the other variable to recover. Keith Price's bibliography (http://iris.usc.edu/vision-notes/bibliography/contents.html) contains many important pointers to the literature and there are additional ones in [48]. The literature on the subject is disparate because studies have addressed wide-ranging aspects of the problem, including problem statements, computational and algorithmic considerations, formal questions, applications, special cases, and enhancement strategies. This disparateness shows in the review [48] which identified, in a repertory of about seventy studies, several distinct constraints used to formulate ego-motion interpretation as an optimization problem and which fan onto an array of about a dozen optimization strategies.

From a conceptual standpoint, ego-motion interpretation can be divided into sharply different categories by the 3D interpretation model used to state the problem, and by distinguishing direct processing from indirect, and dense interpretation from sparse. In each category of this division the optimization scheme is generally dictated by the specificities and assumptions proper to the category. Here, we will discuss dense variational methods. The formulations in these methods minimize an objective functional where the 3D variables of interpretation are referenced over the image domain rather than just at a few points.

6.3.1 Indirect Interpretation

Indirect interpretation of ego-motion assumes optical flow estimated and given as input. In a stationary environment the problem can be addressed using the Longuet-Higgins and Prazdny rigid motion model Eq. (6.6) in a functional of the form:

$$\mathscr{E}(\omega, \mathbf{T}, Z) = \int_{\Omega} \left((u - \xi_u)^2 + (v - \xi_v)^2 + g(\|\nabla Z\|) \right) dxdy, \tag{6.14}$$

where g is a function modifying the norm of the gradient of depth as a means of regularization as discussed in previous chapters, and ξ_u, ξ_v are given by the righthand side of Eq. (6.6):

$$\begin{aligned} \xi_u &= -\frac{xy}{f}\omega_1 + \frac{f^2 + x^2}{f}\omega_2 - y\omega_3 + \frac{f\tau_1 - x\tau_3}{Z} \\ \xi_v &= -\frac{f^2 + y^2}{f}\omega_1 + \frac{xy}{f}\omega_2 + x\omega_3 + \frac{f\tau_2 - y\tau_3}{Z}, \end{aligned}$$

Differentiation under the integral sign with respect to ω and $\mathbf{T}$ and the Euler-Lagrange equations with respect to Z lead to a greedy algorithm which, following initialization of the environment as a frontoparallel plane, i.e., constant depth, iterates two consecutive steps: 3D motion estimation by linear least squares assuming depth fixed, and depth computation by gradient descent assuming motion fixed. The estimation of the translational component of motion and depth is, of course, subject to the uncertainty of scale discussed earlier. The minimization equations are simple to write using the basic formulas in Chap. 2 and we leave it as an exercise to the reader.

The study in [22] used a functional similar to Eq. (6.14) but without regularization, and which it minimized by first solving analytically for depth as a function of motion, and then substituting the solution into the functional and minimizing the resulting depth-free functional via a system of nonlinear equations. This procedure essentially performed $\arg\min_{\mathbf{T},\omega} (\arg\min_Z \mathscr{E}(\mathbf{T}, \omega, Z))$. A modification which used a result from [25] and auxiliary variables to linearize the motion estimation part of this scheme has been investigated in [49].

Rather than using the Longuet-Higgins and Prazdny model, the problem can be expedited using its linearized version Eq. (6.9), i.e., by minimizing, first, the functional:

$$\mathscr{E}(\mathbf{e}) = \int_{\Omega} (< \mathbf{d}, \mathbf{e} >)^2 \, dxdy \tag{6.15}$$

to solve by least squares, up to scale, for the essential parameter vector **e**, from which the rigid motion screw parameters can be recovered uniquely by Eqs. (6.7–6.8), followed by the recovery of depth by the Longuet-Higgins and Prazdny model equations. One can determine **e** simply by writing Eq. (6.9) for every pixel of the digital image and solving the resulting overdetermined homogeneous system of linear equations by least squares using, for instance, the singular value decomposition (SVD) method [6, 50].

6.3.2 Direct Interpretation

Optical flow does not appear in the objective functional of direct interpretation. Such a functional can be written using the 3D rigid motion constraint Eq. (6.12):

$$\mathscr{E}(\omega, \mathbf{T}, Z) = \int_{\Omega} \xi^2 dxdy + \int_{\Omega} g(\|\nabla Z\|) dxdy, \tag{6.16}$$

where $\xi = \xi(x, y)$ is defined in terms of the lefthand side of the 3D rigid body motion constraint model (Eqs. 6.12–6.13):

$$\xi = \frac{1}{Z} < \mathbf{s}, \mathbf{T} > + < \mathbf{q}, \omega > + I_t \tag{6.17}$$

The formulation of [11, 51] was along this vein but applied to a planar scene, with an extension to quadratic patches in [52]. A similar formulation in [12] was applied to the special cases of known depth, pure rotational motion, pure translational motion or rigid motion with known rotational component.

In general, the minimization of Eq. (6.16) can be done as in the indirect formulation by a greedy algorithm which, following an initialization of depth, say to that of a frontoparallel plane, iterates two consecutive steps: minimization with respect to motion by linear least squares assuming depth is fixed, followed by the minimization with respect to depth by gradient descent assuming motion fixed. When Z is considered fixed, the minimization with respect to $\mathbf{T}$, ω reduces to linear least squares because ξ is a linear function of these parameters. Specifically, let

$$\rho = (\mathbf{T}, \omega)^T = (\tau_1, \tau_2, \tau_3, \omega_1, \omega_2, \omega_3)^T, \tag{6.18}$$

and

$$\mathbf{a} = \left(q_1, q_2, q_3, \frac{s_1}{Z}, \frac{s_2}{Z}, \frac{s_3}{Z}\right), \tag{6.19}$$

where s_1, s_2, s_3 are the coordinates of $\mathbf{s}$ defined in Eq. (6.13) and q_1, q_2, q_3 are those of $\mathbf{q}$. By differentiation of the objective functional with respect to the rigid motion parameters $\rho_j,\ j = 1, \ldots, 6$, we obtain the normal equations of minimization:

$$< \mathbf{b}_j, \rho > = \mathbf{r}_j \quad j = 1, \ldots, 6, \tag{6.20}$$

where $\mathbf{r}_j$ is

$$\mathbf{r}_j = -\int_{\Omega} a_j I_t \, dxdy, \tag{6.21}$$

and component i of $\mathbf{b}_j$ is:

$$\mathbf{b}_j^i = \int_{\Omega} a_j a_i \, dxdy, \quad i = 1, \ldots, 6. \tag{6.22}$$

With digital images, the estimation can be expedited by setting up an overdetermined system of linear equations as follows and solving it by an efficient routine such as the SVD method. Writing the 3D rigid body constraint Eq. (6.12) for each $\mathbf{x}_i$ we obtain an overdetermined system of linear equations:

$$\mathbf{A}\,\rho = \mathbf{c}, \tag{6.23}$$

where $\mathbf{A}$ and $\mathbf{c}$ are defined by:

$$\mathbf{A} = \begin{bmatrix} \mathbf{a}(\mathbf{x}_1) \\ \vdots \\ \mathbf{a}(\mathbf{x}_n) \end{bmatrix} \quad \mathbf{c} = \begin{bmatrix} -I_t(\mathbf{x}_1) \\ \vdots \\ -I_t(\mathbf{x}_n) \end{bmatrix},$$

with n being the number of pixels. This overdetermined homogeneous linear system is then solved up to scale for least squares 3D motion, by the SVD method for instance.

6.4 Non-Stationary Environment

We will now study the case where environmental objects and the viewing system can move simultaneously and independently. We will assume that the objects are rigid. The purpose of 3D interpretation would be to recover 3D structure and motion of each object that moves *relative* to the viewing system. Therefore, by motion we will mean motion relative to the viewing system, that the viewing system moves or not. Three-dimensional interpretation is now faced with the problem of recovering not the single viewing system motion as in ego-motion in a static environment but, instead, the relative motion of each object moving independently. This complicates the problem significantly because 3D interpretation is now tied to the notion of *motion-based image segmentation*, where regions correspond to differently moving objects in space, relative to the viewing system, or to the dual notion of motion boundaries, which correspond to transitions in the image from one motion to another significantly different. This is so because the 3D motion of an object is to be estimated with the image data corresponding to the projection of the object which, therefore, must be delineated. This delineation can be done by *region-based* interpretation and explicit 3D motion region processing, or by *point-wise* interpretation via motion boundary preserving regularization of the type we have discussed for optical flow estimation in Chap. 3, via the Aubert function for instance [53]. We will treat both cases in the following sections.

6.4.1 Point-Wise Interpretation

Point-wise interpretation has been addressed in [38, 54] and [55]. It views motion and depth as functions of image position, i.e., are allowed to vary from point to point. The 3D interpretation is computed by minimizing an objective functional which contains a data term of conformity of the interpretation to the image spatiotemporal variations and a regularization term which preserves 3D motion boundaries so as to account for the various differently moving objects in the viewed environment.

The direct scheme of interpretation in [38, 54] uses the rigid 3D motion constraint Eq. (6.12) to define the data term. It also uses a regularization that preserves 3D interpretation discontinuities. More, precisely, let $\mathbf{T}_Z = \mathbf{T}/Z; \quad \rho = (\omega, \mathbf{T}_Z)$, and $\mathbf{r} = (\mathbf{q}, \mathbf{s})$. With this notation, Eq. (6.12) is rewritten:

$$\mathbf{r} \cdot \rho + I_t = 0. \tag{6.24}$$

The 3D interpretation is sought by minimizing the following functional:

$$\mathscr{E}(\rho) = \int_\Omega (\mathbf{r} \cdot \rho + I_t)^2 + \lambda \sum_{j=1}^{6} \int_\Omega g(\|\nabla \rho_j\|)\, dx\, dy \tag{6.25}$$

The Euler-Lagrange equations corresponding to the minimization of this functional are:

$$\lambda\, div\left(\frac{g'(\|\nabla \rho_j\|)}{\|\nabla \rho_j\|} \nabla \rho_j\right) = 2 r_j (\mathbf{r} \cdot \rho + I_t) \qquad j = 1, \ldots, 6 \tag{6.26}$$

A discretization of Eq.(6.26) gives a large system of nonlinear equations. Rather than solving this system, the study [38, 54] minimized the objective functional using the half-quadratic algorithm of [53, 56] which we described in Chap. 3. Functional Eq. (6.25) is minimized via the minimization of the following other functional defined through a vector field of auxiliary variables $\mathbf{b} = (b_1, b_2, \ldots, b_6)$:

$$\mathscr{E}^*(\rho, \mathbf{b}) = \int_\Omega (\mathbf{r} \cdot \rho + I_t)^2 + \lambda C^*(\rho, \mathbf{b})\ dx\, dy, \tag{6.27}$$

where

$$C^*(\rho, \mathbf{b}) = \sum_{j=1}^{6} \left(b_j \|\nabla \rho_j\|^2 + \psi(b_j)\right), \tag{6.28}$$

and ψ is a strictly decreasing convex function implicitly related to g and the explicit expression of which is not needed by the algorithm execution. Details of the implementation and concomitant discretization are given in [38].

A direct method along the vein of [38, 54] has been investigated in [55] with a generalization of the data term as an approximate L^1 metric of the displaced frame difference, i.e., the data term is:

$$\mathscr{E}_D(\rho) = \int_{\Omega} g\big(I(x + u(\rho), y + v(\rho), t + 1) - I(x, y, t)\big) dxdxy, \tag{6.29}$$

where $g = \sqrt{z^2 + \varepsilon^2}$ for small ε realizes an approximate L^1 metric, and the image displacements (u, v) between views are given linearly in terms of the 3D variables of interpretation $\rho = (\mathbf{T}_Z, \omega)$:

$$u(\rho) = -\frac{xy}{f}\rho_1 + \frac{f^2 + x^2}{f}\rho_2 - y\rho_3 + f\rho_4 - x\rho_6 \tag{6.30}$$

$$v(\rho) = -\frac{f^2 + y^2}{f}\rho_1 + \frac{xy}{f}\rho_2 + x\rho_3 + f\rho_5 - y\rho_6. \tag{6.31}$$

Details of a multiresolution implementation of the Euler-Lagrange equations corresponding to Eq. (6.29) are given in [55].

Once ρ is computed, depth can be recovered when $\mathbf{T} \neq \mathbf{0}$ from $\mathbf{T}_Z = \mathbf{T}/Z = (\rho_1, \rho_2, \rho_3)$ by:

$$\frac{1}{Z} = \sqrt{\rho_1^2 + \rho_2^2 + \rho_3^2}, \tag{6.32}$$

which effectively amounts to resolving the uncertainty of scale by imposing unit length translational velocity $\|\mathbf{T}\| = 1$, i.e., recovering the direction of the translational velocity vector.

The methods in [38, 54] and [55] are direct insomuch as they substitute a parametric model of optical flow in terms of 3D variables directly in the objective functional. Using constraint Eq. (6.9), rather than Eq. (6.12), one can formulate an indirect scheme, where optical flow is used as data, by minimizing an objective functional of the form:

$$\mathscr{E}(\mathbf{e}) = \int_{\Omega} (<\mathbf{d}, \mathbf{e}>)^2 \; dxdy + \lambda \sum_{j=1}^{6} \int_{\Omega} g(\|\nabla e_j\|) \; dxdy \tag{6.33}$$

The corresponding Euler-Lagrange equations, subject to the uncertainty of scale, are given by:

$$2d_j < \mathbf{d}, \mathbf{e} > +\lambda \, div\left(\frac{g'(\|\nabla e_j\|)}{\|\nabla e_j\|}\nabla e_j\right) = 0 \qquad j = 1, \ldots, 6 \tag{6.34}$$

Algorithms as in [38] and [55] can be used to solve these equations.

Example: Here following is an example for the purpose of showing what kind of results one can expect with point-wise direct 3D interpretation. The results shown

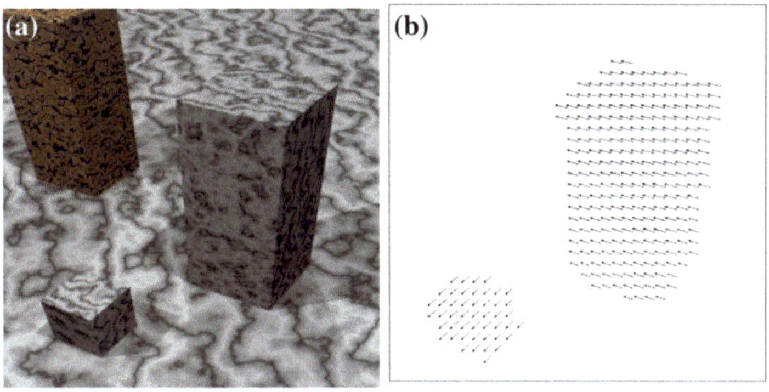

Fig. 6.2 Point-wise direct interpretation: **a** the first of the two consecutive images of the *Marbled blocks* sequence used and, **b** the ground truth optical flow between the two views

were produced by the method of [38]. The example uses the *Marbled block* synthetic sequence (KOGS/IAKS laboratory image database) which consists of images of two moving blocks in a static environment. The first of the two consecutive images used is displayed in Fig. 6.2a. The larger block moves to the left in depth and the smaller forward to the left. Figure. 6.2b shows the ground truth displacements between the two views used (KOGS/IAKS laboratory image database).

The *Marbled block* sequence is interesting because it has aspects that challenge 3D reconstruction. First the blocks texture is composed of weak intensity contrast textons. At the top, and particularly for the larger block, this texture is the same as of the background, causing the corresponding intensity boundaries to be faint and, therefore, ambiguous. Second, the blocks sides hidden from the light source are significantly shadowed. Finally, there are occlusion boundaries with sharp depth discontinuities.

The focal length was set to $f = 600$ pixels. This corresponds approximately to the focal length of an 8.5 mm camera with inter-pixel distance of 0.015 mm. Varying the focal length about 600 pixels did not have a noticeable effect on the recovered structure, a behaviour that is consistent with a remark in [57] citing the literature on self calibration [58].

The recovered structure is displayed by grey level rendering and by an anaglyph. Anaglyphs are generated using a stereoscopic image constructed from the first image used and the estimated depth [59]. Depth computed from the 3D interpretation is shown in Fig. 6.3a and a corresponding anaglyph in Fig. 6.3b. Note that the boundary-preserving regularization of the 3D interpretation has preserved the occluding edges of each block as well as the edges between their visible facets. However, the faces are not as flat as they should be probably due to the presence of patches of significantly weak intensity contrast where depth could not be sufficiently corrected by the point-wise process of regularization. The image motion reconstructed from the 3D interpretation is displayed in Fig. 6.4a and the Horn and Schunck optical flow is shown in Fig. 6.4b, showing that the algorithm has done a decent job at recovering 3D motion.

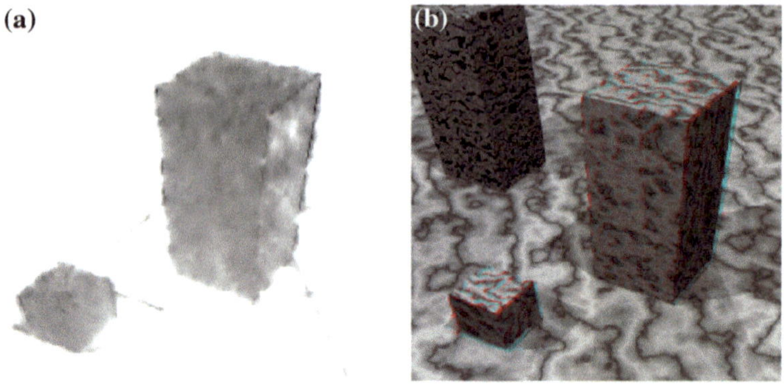

Fig. 6.3 Point-wise direct interpretation: **a** *Grey* level rendering of the estimated depth and, **b** an anaglyph of the scene, to be viewed with *red-cyan* glasses

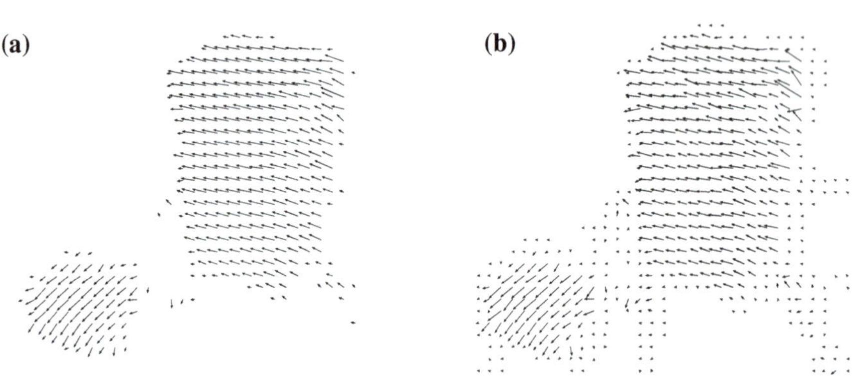

Fig. 6.4 Point-wise direct interpretation: **a** Optical flow estimated from the 3D interpretation and **b** optical flow by the Horn and Schunck method

6.4.2 Region-Based Interpretation

Region-based interpretation will reference and seek to determine maximal regions corresponding to the moving environmental objects which, as in the previous discussions, we will assume are rigid. This amounts to image segmentation according to the movement of real objects, which necessarily brings in the 3D interpretation formulation variables related to the motion and structure of the objects. Then, obviously, segmentation and 3D motion and structure recovery are interdependent processes: the segmentation needs the 3D variables of each object region and the estimation of the objects 3D variables must be performed exclusively within each object region. Therefore, it would be advantageous to perform segmentation and recovery of 3D variables concurrently.

Active contours and the level set representation have been quite effective in image segmentation at large [20, 60] and we have seen in the preceding chapters that they can be productive tools in optical flow estimation, and motion detection and tracking. For 3D interpretation as well, we will show in the subsequent sections that they can afford efficient algorithms by allowing encoding of 3D interpretation, 3D-motion segmentation, and optical flow estimation in a single objective functional which yields to effective optimization.

Under the assumption that moving environmental objects are rigid, region-based 3D interpretation methods can use, as with point-wise processing methods, the Longuet-Higgins and Prazdny optical flow rigid body model Eq. (6.6), or its linearized expression Eq. (6.9), or the Negahdaripour 3D rigid motion constraint Eq. (6.12), to construct a data term which evaluates the fidelity of the 3D variables to the image data within each segmentation region. Image data is either optical flow evaluated beforehand or concurrently with the 3D variables, or the image sequence spatiotemporal variations. In active contour formulations, regions are defined from closed simple plane curves as we have seen many times in preceding chapters.

Here following are region-based formulations of active contour/level set methods of 3D interpretation and segmentation of image sequences: (1) a *3D rigid motion constraint formulation* which uses a data fidelity term based on Eq. (6.12) to state joint 3D-motion segmentation and estimation of depth and 3D motion and, (2) a *depth-free formulation* where depth is eliminated from the objective functional via the use of a data term based on Eq. (6.9), which brings in 3D rigid motion essentials parameters and optical flow in the problem statement but not depth. Optical flow does not appear in the objective functional of the first formulation, which implies that the method is direct. Optical flow appears in the second formulation, which means that both direct and indirect versions can be considered.

For a clearer presentation of the formulations, we will treat the case of two-region segmentation before multiregion partitioning.

Let $I : (x, y, t) \in \Omega \times]0, T[\mapsto I(x, y, t) \in \mathbf{R}^+$ be an image sequence with common domain Ω and duration T, possibly acquired by a moving viewing system. We will investigate the problem of dividing Ω into two regions, R_1 and $R_2 = R_1^c$, on the basis of 3D motion, and for each region determine the corresponding 3D structure and motion. For the purpose of describing the boundary of R_1, let $R_1 = R_\gamma$, where γ is a closed simple plane curve and R_γ its interior. Then $R_2 = R_\gamma^c$.

6.4.2.1 3D Rigid Motion Constraint Formulation

Let Z designate the depth function over Ω and $\omega_k, \mathbf{T}_k,\ k = 1, 2$ the rigid motion parameter vectors assigned respectively to R_1 and R_2. Consider the following objective functional of direct interpretation [39] which involves all of the unknowns, i.e., depth, the parameters of the screws of 3D rigid motion, and the active curve γ:

$$\mathscr{E}\left(\gamma, \{\omega_k, \mathbf{T}_k\}_{k=1}^2, Z\right) = \sum_{k=1}^{2} \int_{R_k} \left(\xi_k^2 + \mu \|\nabla Z\|^2\right) dxdy + \lambda \oint_\gamma ds, \tag{6.35}$$

where μ and λ are positive real constants to modulate the contribution of the terms they multiply, and the ξ_k's are given by the lefthand side of the 3D rigid motion constraint Eq. (6.12):

$$\xi_k = \mathbf{s} \cdot \frac{\mathbf{T}_k}{Z} + \mathbf{q} \cdot \omega_k + I_t, \tag{6.36}$$

with data vectors **s** and **q** defined in Eq. (6.13). For each region R_k, $k = 1, 2$, the first of the two terms in the first integral of Eq. (6.35), ξ_k^2, is a data function of image position to evaluate the conformity of the 3D motion parameters ω_k, $\mathbf{T}_k$ to the image sequence spatiotemporal derivatives via the Negahdaripour 3D rigid motion constraint. The other function regularizes depth by smoothness within each region. The second integral is the usual length term to bias the segmentation to have a smooth boundary γ.

Minimizing the functional with respect to all its arguments, namely, depth, the 3D motion parameters within each of the two regions R_1 and R_2, and curve γ which defines these regions, will partition the image into two regions separated by a smooth boundary and give for each the smooth rigid structure and corresponding motion that best explain the image brightness spatiotemporal variations, namely, its spatiotemporal derivatives, in conformity with the 3D rigid body constraint Eq. (6.12) and the viewing system model depicted in Fig. 6.1.

The minimization of Eq. (6.35) can be done by a greedy scheme which iterates three consecutive steps until convergence, A. computation of motions parameters considering the curve and depth fixed, B. computation of depth with the curve and motion fixed and, C. curve evolution with depth and motion fixed. The initialization starts the process with a curve, say a circular contour placed so as to cover about half the image with its interior, and the constant depth of a frontoparallel plane. The three steps of the algorithm are as follows, in the order of instantiation:

A. *Update of motion*:

With Z and γ fixed, i.e., taken for known and used as data at this step, the energy to minimize is:

$$\mathscr{E}_{Motion}\left(\{\mathbf{T}_k, \omega_k\}_{k=1}^2\right) = \sum_{k=1}^{2} \int_{R_k} \xi_k^2 \, dxdy \tag{6.37}$$

Each ξ_k being linear in $\mathbf{T}_k$, ω_j, the minimization amounts to linear least-squares estimation of the parameters in each region. For a digital image, this is done by solving, say by the SVD method, an overdetermined system of linear equations for each of the two current regions $R_1 = R_\gamma$ and $R_2 = R_\gamma^c$:

$$\mathbf{A}_k \, \rho_k = \mathbf{c}_k, \quad k = 1, 2 \tag{6.38}$$

where $\rho_k = (\omega_k, \mathbf{T}_k)^T$, matrix $\mathbf{A_k}$ and vector $\mathbf{c}_k$ are constructed from pixels $\mathbf{x}_1, \ldots, \mathbf{x}_{\mathbf{n}_k}$ of region R_j:

$$\mathbf{A}_k = \begin{bmatrix} \mathbf{a}(\mathbf{x}_1) \\ \vdots \\ \mathbf{a}(\mathbf{x}_{n_k}) \end{bmatrix} \quad \mathbf{c}_k = \begin{bmatrix} -I_t(\mathbf{x}_1) \\ \vdots \\ -I_t(\mathbf{x}_{n_k}) \end{bmatrix},$$

and the expression of **a** is given in Eq. (6.19).

B. *Update of depth*:

Taking the current motion and curve to be fixed and useable as data, depth is updated by minimizing:

$$\mathcal{E}_{Depth}(Z) = \sum_{k=1}^{2} \int_{R_j} \left[\xi_k^2 + \mu g(\|\nabla Z\|)\right] dxdy \tag{6.39}$$

$$= \int_{\Omega} \sum_{k=1}^{2} \chi_k \left[\xi_k^2 + \mu g(\|\nabla Z\|)\right] dxdy, \tag{6.40}$$

where χ_k is the characteristic function of region R_k and g is a function to preserve boundaries. The corresponding Euler-Lagrange descent equation to update depth is:

$$\frac{\partial Z}{\partial \tau} = -\frac{\partial \mathcal{E}_{Depth}}{\partial Z} = \sum_{k=1}^{2} \chi_k \left[2\frac{\mathbf{s} \cdot \mathbf{T}_k}{Z^2}\xi_k + \mu \operatorname{div}\left(\frac{g'(\|\nabla Z\|)}{\|\nabla Z\|}\nabla Z\right)\right], \tag{6.41}$$

where τ is the algorithmic time. Rather than using a discontinuity-preserving function, one can expedite processing by using the quadratic function $g(z) = z^2$ in which case the divergence term in Eq. (6.41) becomes the Laplacian of depth, $\nabla^2 Z$, and evaluate this Laplacian along the boundary curve γ according to an ad hoc discontinuity-preserving approximation [61].

C. *Update evolution of* γ:

To evolve $\boldsymbol{\gamma} : [0, 1] \to \Omega$, we embed it in a one-parameter family $\boldsymbol{\gamma} : [0, 1] \times R^+ \to \Omega$ of closed regular curves indexed by algorithmic time τ and move it according to the Euler-Lagrange descent equation $\frac{d\boldsymbol{\gamma}}{dt} = -\frac{\delta \mathcal{E}}{\delta \boldsymbol{\gamma}}$ (refer to Chap. 2). Assuming both current motion and depth are fixed at this step, useable as data, this equation is:

$$\frac{d\gamma}{d\tau} = -\frac{\partial \mathcal{E}}{\partial \gamma} = -\left(\eta_1 - \eta_2 + \lambda\kappa\right)\mathbf{n}, \tag{6.42}$$

where **n** is the outward unit normal function of γ and κ its curvature function, and η_1, η_2 are given by:

$$\eta_k = \xi_k^2 + \mu g(\|\nabla Z\|), \quad k = 1, 2 \tag{6.43}$$

The corresponding level set evolution equation (see Chap. 2) is:

$$\frac{d\phi}{d\tau} = -(\eta_1 - \eta_2 + \lambda\kappa)\,\|\nabla\phi\|, \tag{6.44}$$

where ϕ is the level set function of which γ is the zero level, positive inside γ and negative outside. Curvature is written in terms of the level set function as:

$$\kappa = \operatorname{div}\left(\frac{\nabla\phi}{\|\nabla\phi\|}\right), \tag{6.45}$$

where $\mathbf{n}$ is oriented outward:

$$\mathbf{n} = -\frac{\nabla\phi}{\|\nabla\phi\|} \tag{6.46}$$

Interpretation up to Scale

Let $\{R_1 = R_\gamma, R_2 = R_\gamma^c\}$ be the segmentation at convergence and $(\{\omega_k, \mathbf{T}_k\}_{k=1}^2, Z)$ the corresponding 3D interpretation. This interpretation satisfies the normal equations Eq. (6.20) in each region. Any interpretation $(\{\omega_k, \alpha\mathbf{T}_k\}_{k=1}^2, \alpha Z)$, $\alpha \in \mathbb{R}^+$ in the same segmentation also satisfies them. Therefore, only the direction of the translational component of 3D motion, and relative depth thereof, can be determined.

Multiregion Extension

Multiregion segmentation, or image partitioning into N regions, $N > 2$, requires at least two active curves. The functional data term can be written as a sum of the regions individual data terms:

$$\mathscr{D} = \sum_{k=1}^{N} \int_{R_k} \psi_k(x, y)\, dxdy, \tag{6.47}$$

where regions $\{R_k\}_1^N$ are defined from the active curves. This definition must ensure that the objective functional minimization yields a set of regions that form a partition, i.e., cover the image domain Ω and do not overlap. Various definitions of regions from closed regular plane curves which result in partitions have been reviewed briefly in Chap. 3 and in more detail in [21].

Example: This is an example of the type of 3D interpretation and motion segmentation results which can be obtained by minimizing the 3D rigid motion constraint functional Eq. (6.35). The scene in Fig. 6.5 is the same as in the preceding example. The goal of segmentation here is to determine three distinct regions of 3D motion, two of them corresponding to the two moving blocks and the third to their back-

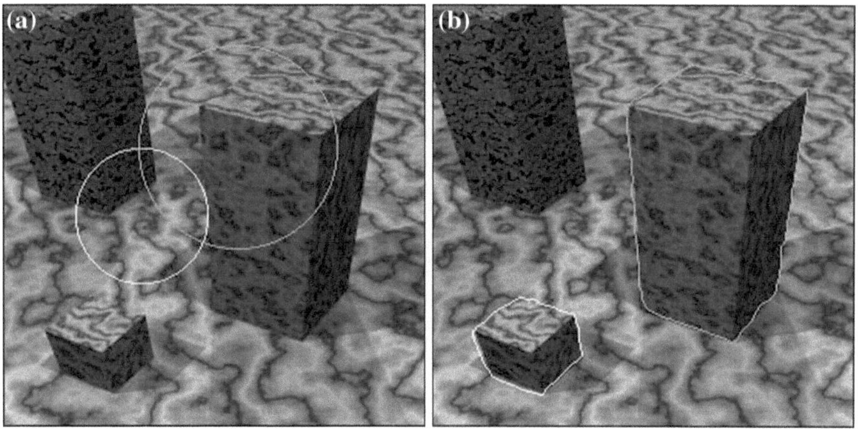

Fig. 6.5 Region-based direct 3D interpretation and 3D motion segmentation by minimizing the 3D rigid motion functional Eq. (6.35): **a** the first of the two consecutive images used in the experiment and the initial *curves*; **b** the final 3D motion segmentation. The final *curves* have closely adhered to the two moving blocks

ground. Therefore, we use two active curves γ_1 and γ_2. The initial position of these two curves is shown in Fig. 6.5a and their final position, produced by the algorithm at convergence, is displayed in Fig. 6.5b. Both moving blocks have been delineated correctly. The reconstructed depth of the blocks, triangulated and shaded according to a local light source, is shown in Fig. 6.6. The structure of both blocks has been correctly determined. The recovered structures have sharply delineated occluding boundaries thanks to the fact that the segmentation process has correctly delineated them.

6.4.2.2 Depth-Free Formulation

Depth can be eliminated from the 3D interpretation formulation by using a data fidelity term based on the homogeneous linear constraint Eq. (6.9) rather than the Negahdaripour rigid 3D motion constraint Eq. (6.12) as in the previous objective function Eq. (6.35). This will bring optical flow in the formulation functional and, as a result, indirect and direct statements of the problem can be considered.

Indirect Formulation

If optical flow is known beforehand, joint 3D motion segmentation and 3D interpretation can be done by minimizing the following functional, in the case of two regions:

(a) (b)

Fig. 6.6 Region-based direct 3D interpretation and 3D motion segmentation by minimizing the 3D rigid motion constraint functional Eq. (6.35): **a, b** triangulation-based surface rendering of the recovered blocks structure. The recovered structures have sharp occluding boundaries due to the fact that the segmentation process has correctly delineated them

$$\mathcal{E}\left(\gamma, \{\mathbf{e}_k\}_{k=1}^{2}\right) = \sum_{k=1}^{2} \int_{\mathbf{R}_k} (\mathbf{d} \cdot \mathbf{e}_k)^2 + \lambda \int_{\gamma} ds, \tag{6.48}$$

where, as before, $R_1 = R_\gamma$ and $R_2 = R_1^c$. The first term evaluates the conformity of the 3D motions, via their essential parameters, to the data, namely optical flow.

The minimization of this functional can be done by a greedy descent which iterates two steps until convergence, computation of the essential parameters in each region with γ fixed, and evolution of γ with the motion parameters fixed. The initialization places a starting curve in Ω.

A. *Essential parameters update*:

The minimization equations with respect to the essential parameters comes to a least squares expression of these in each of the two current regions R_γ and R_γ^c. Computations consist of solving, in each region, an overdetermined system of linear equations, by the singular value method, for instance:

$$\mathbf{D}_k \mathbf{e}_k = 0, \quad k = 1, 2, \tag{6.49}$$

where, for $k = 1, 2$, the rows of matrix $\mathbf{D}_k$ correspond to the data vector $\mathbf{d}$ defined by Eq. (6.10) evaluated at the points of region R_k. The equations are homogeneous. The solution is obtained up to a scale factor, for instance the unit norm solution: $\|\mathbf{e}_k\| = 1$.

B. *Curve evolution*:

Curve γ is embedded in a one-parameter family $\boldsymbol{\gamma} : [0, 1] \times R^+ \rightarrow \Omega$ of closed regular curves indexed by algorithmic time τ and evolved according to the Euler-Lagrange descent equation:

$$\frac{d\gamma}{d\tau} = -\frac{\delta \mathscr{E}}{\delta \gamma} = -\big((<\mathbf{d}, \mathbf{e}_1>)^2 - (<\mathbf{d}, \mathbf{e}_2>)^2 + \lambda\kappa\big)\mathbf{n}, \tag{6.50}$$

where $\mathbf{e}_1, \mathbf{e}_2$ are the essential parameters obtained at the parameter update step and considered fixed for this step of curve evolution. The corresponding level set equation is:

$$\frac{d\phi}{d\tau} = -\left((<\mathbf{d}, \mathbf{e}_1>)^2 - (<\mathbf{d}, \mathbf{e}_2>)^2 + \lambda\kappa\right) \|\nabla\phi\| \tag{6.51}$$

Direct Formulation

When optical flow is not known beforehand, it can still be estimated concurrently with the segmentation and the essential parameters by minimizing a functional which includes terms for its estimation [40]:

$$\begin{aligned}
\mathscr{E}\left(\gamma, \{\mathbf{e}_k\}_{k=1}^2, u, v\right) &= \sum_{k=1}^{2} \int_{R_k} (<\mathbf{d}, \mathbf{e}_k>)^2 \, dxdy \\
&+ \sum_{k=1}^{2} \int_{R_k} \mu(\nabla I \cdot \mathbf{w} + I_t)^2 + \nu\big(g(\|\nabla u\|) + g(\|\nabla v\|)\big) \, dxdy \\
&+ \lambda \oint_{\gamma} ds
\end{aligned} \tag{6.52}$$

The terms for the estimation of optical flow are in the middle line of Eq. (6.52). These are common terms, namely a data term of optical flow conformity to the image spatiotemporal derivatives and terms of smoothness via a regularization function g. The minimization of Eq. (6.52) can be done by iterative greedy descent here also, but with three consecutive steps instead of two as for Eq. (6.48): optical flow computation, essential parameter computation, and curve evolution. The last two steps are identical to those of the minimization of Eq. (6.48): essential parameter vectors update governed by Eq. (6.49) assuming γ and optical flow u, v fixed, and evolution of γ, governed by Eq. (6.50), assuming the motion parameters and optical flow fixed.

For the first step, the Euler-Lagrange equations corresponding to the minimization with respect to optical flow within each region $R_k, k = 1, 2$, assuming γ and the motion parameters fixed, are:

$$
\begin{aligned}
(fe_{k,8} - ye_{k,9}) < \mathbf{d}, \mathbf{e}_k > +\mu I_x(\nabla I \cdot \mathbf{w} + I_t) - \nu \operatorname{div}\left(\frac{g'(|\nabla u|)}{|\nabla u|}\nabla u\right) &= 0 \\
(-fe_{k,7} + xe_{k,9}) < \mathbf{d}, \mathbf{e}_k > +\mu I_y(\nabla I \cdot \mathbf{w} + I_t) - \nu \operatorname{div}\left(\frac{g'(|\nabla v|)}{|\nabla v|}\nabla v\right) &= 0
\end{aligned}
\tag{6.53}
$$

If one assumes that optical flow boundaries occur only at region boundaries, i.e., along γ, then the flow estimation can be expedited by using the quadratic function $g(z) = z^2$, in which case the divergence term is the Laplacian and Eq. (6.53) results in a large sparse system of linear equations which can be solved efficiently by iterative methods [62, 63].

Recovery of Relative Depth

The translational and rotational components of rigid 3D motion can be recovered analytically and uniquely from the essential parameters, for each region separately. The translational component $\mathbf{T} = (t_1, t_2, t_3)$ is given by Eq. (6.8) up to a sign and a positive scale factor: $t_1 = e_7, t_2 = e_8, t_3 = e_9$. The rotational component is computed from Eq. (6.7). When $\mathbf{T} \neq \mathbf{0}$, depth can be pulled out of the Longuet-Higgins and Pradzny rigid motion equations (6.6) and computed as:

$$
Z = \sqrt{\frac{(ft_1 - xt_3)^2 + (ft_2 - yt_3)^2}{\left(u + \frac{xy}{f}\omega_1 - \frac{f^2+x^2}{f}\omega_2 + y\omega_3\right)^2 + \left(v + \frac{f^2+y^2}{f}\omega_1 - \frac{xy}{f}\omega_2 - x\omega_3\right)^2}}.
\tag{6.54}
$$

Since the components of $\mathbf{T}$ appear in a ratio with depth in the Longuet-Higgins and Prazdny model and translation is recovered up to a scale factor, only relative depth up to the same scale factor is recovered. This factor is determined when the essential parameters are computed in each region under a fixed norm constraint. Once depth is computed, the sign of $\mathbf{T}$ is adjusted if necessary to correspond to positive depth [6, 15].

The fact that only the direction of translation, and relative depth thereof, can be recovered implies two things: (1) 3D motions with the same rotational component and the same direction of translational components cannot be distinguished, although this does not affect the recovery of depth and, (2) the depth in one segmented region is relative to the depth in another in the ratio of the norm of their actual translational components of motion.

Multiregion Extension

Multiregion segmentation into N regions, $N > 2$, is done exactly as we described it for the preceding method: The functional data term can be written as a sum of the regions individual data terms:

$$\mathscr{D} = \sum_{i=1}^{N} \int_{R_i} \psi_i(x, y)\, dxdy, \tag{6.55}$$

where regions $\{R_i\}_1^N$ are defined from the active curves. We again refer the reader to the review in Chap. 3 and to [21] for more details.

Example: This example illustrates the type of direct 3D interpretation results that can be obtained by minimizing the region-based, depth-free functional Eq. (6.52). The scene, shown in Fig. 6.7a, contains three moving real objects (courtesy of Debrunner and Ahuja [64]). There is a cylindrical surface moving laterally to the right at an image velocity of about 0.15 pixel per frame and also rotating approximately one degree per frame about its axis. There also is a box moving to the right at about 0.30 pixel per frame and a flat background moving right at approximately 0.15 pixel per frame. For the purpose of 3D-motion segmentation, the box and background are considered a single object because they move according to parallel translations (for which only the direction can be recovered as we have seen). An anaglyph of the recovered scene is depicted in Fig. 6.7b. Figs. 6.8a and b show the recovered structure of the cylindrical object and of the box-and-background as projections of their triangulated surfaces wrapped with the original image.

Next, we will estimate scene flow as a direct means of describing 3D motion without assuming that moving environmental objects are rigid.

6.5 Scene Flow

Scene flow is the 3D velocity field of the visible environmental points, i.e., it is the 3D vector field over the image domain which consists at each point of the velocity of the corresponding surface point in space. Using the notation in Eq. (6.3), it is the field (U, V, W) over the image domain Ω.

Scene flow can be recovered by methods such as those we have described in the preceding sections, which use a monocular image sequence and a parametric form for the flow, for instance by interpreting the flow to be the velocity of a rigid body as shown in Eq. (6.5). The constraints on the flow are then the constraints on its

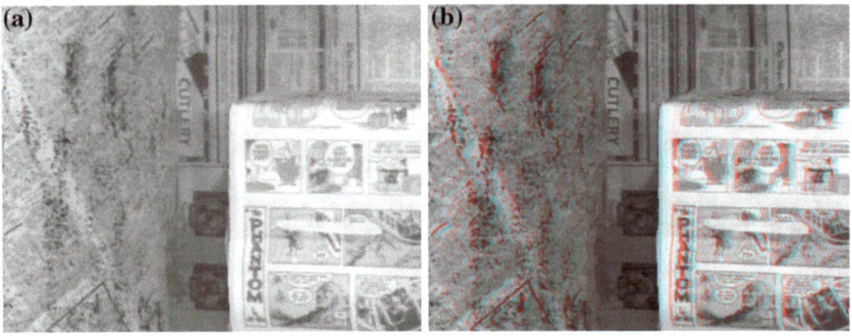

Fig. 6.7 Direct 3D interpretation and motion segmentation by minimizing the region-based, depth-free functional Eq. (6.52): **a** the scene contains three moving objects: to the *left* there is a cylindrical surface moving laterally to the *right* and rotating about its axis; to the *right* there is a box moving to the *right*; and in the *back* there is a flat background also moving *right*. For the purpose of 3D-motion segmentation, the box and background are considered a single object because they move according to parallel translations, for which only the direction can be recovered; **b** an anaglyph of the recovered scene (to be viewed with *red/blue* glasses)

representation parameters and the flow is recovered a posteriori from its parametric expression.

Because scene flow is related to optical flow and depth, as Eq. (6.3) in monocular imaging reveals, nonparametric scene flow recovery has been investigated in the context of stereoscopy [43–47], in which case constraints on depth and optical flow by correspondence become available [65, 66]. However, nonparametric scene flow estimation can be stated using a monocular image sequence, i.e., without requiring stereoscopy. A linear formulation is as follows [67]:

By substitution of Eq. (6.3) in the Horn and Schunck optical flow constraint Eq. (6.11) and multiplication by Z give the following linear equation in the variables of scene flow and depth:

$$fI_xU + fI_yV - (xI_x + yI_y)W + I_tZ = 0 \tag{6.56}$$

Multiplication of motion and depth by the same constant maintains the equation. One can remove this uncertainty of scale by choosing depth to be *relative* to the frontoparallel plane $Z = Z_0 > 0$ for some $Z_0 > f$, giving the equation:

$$fI_xU + fI_yV - (xI_x + yI_y)W + I_t(Z - Z_0) + I_tZ_0 = 0 \tag{6.57}$$

For notational simplicity and economy, we reuse the symbol Z to designate relative depth, i.e., relative to plane $Z = Z_0$, and write Eq. (6.57) as:

$$fI_xU + fI_yV - (xI_x + yI_y)W + I_tZ + I_tZ_0 = 0 \tag{6.58}$$

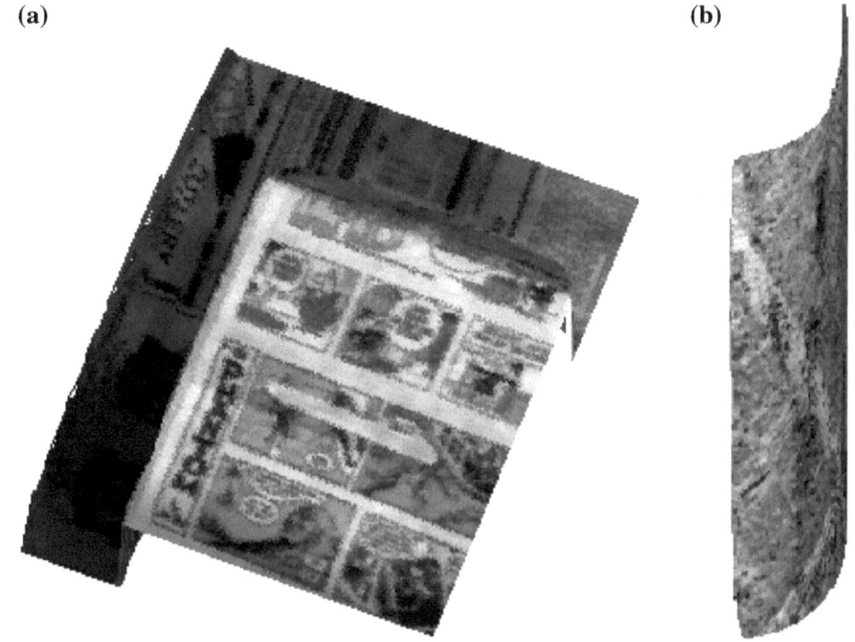

Fig. 6.8 Direct 3D interpretation and motion segmentation by minimizing the region-based, depth-free functional Eq. (6.52): **a** The recovered structure of the box-and-background. The box and the background planar surfaces are taken as a single object because they have parallel translations; **b** the recovered structure of the cylindrical object. The displays are projections of the recovered objects triangulated surfaces wrapped with the original image

Scene flow and relative depth can now be estimated by minimizing the following functional [67]:

$$\mathcal{E}(U,V,W,Z|I) = \frac{1}{2}\int_\Omega (fI_xU + fI_yV - (xI_x + yI_y)W + I_tZ + I_tZ_0)^2 dxdy + \frac{\lambda}{2}\int_\Omega (g(\|\nabla U\|) + g(\|\nabla V\|) + g(\|\nabla W\|) + g(\|\nabla Z\|))dxdy, \tag{6.59}$$

where λ is a positive constant to weigh the relative contribution of the two terms of the functional. With the L^2 regularization, $g(z) = z^2$, we have:

$$\mathcal{E}(U,V,W,Z|I) = \frac{1}{2}\int_\Omega (fI_xU + fI_yV - (xI_x + yI_y)W + I_tZ + I_tZ_0)^2 dxdy + \frac{\lambda}{2}\int_\Omega (\|\nabla U\|^2 + \|\nabla V\|^2 + \|\nabla W\|^2 + \|\nabla Z\|^2)dxdy \tag{6.60}$$

The corresponding Euler-Lagrange equations are:

$$
\begin{aligned}
&fI_x(fI_xU + fI_yV + (-xI_x - yI_y)W + I_tZ + I_tZ_0) - \lambda\nabla^2U = 0\\
&fI_y(fI_xU + fI_yV + (-xI_x - yI_y)W + I_tZ + I_tZ_0) - \lambda\nabla^2V = 0\\
&(-xI_x - yI_y)(fI_xU + fI_yV + (-xI_x - yI_y)W + I_tZ + I_tZ_0) - \lambda\nabla^2W = 0\\
&I_t(fI_xU + fI_yV + (-xI_x - yI_y)W + I_tZ + I_tZ_0) - \lambda\nabla^2Z = 0,
\end{aligned}
\tag{6.61}
$$

with the Neumann boundary conditions:

$$
\frac{\partial U}{\partial \mathbf{n}} = 0, \quad \frac{\partial V}{\partial \mathbf{n}} = 0, \quad \frac{\partial W}{\partial \mathbf{n}} = 0, \quad \frac{\partial Z}{\partial \mathbf{n}} = 0, \tag{6.62}
$$

where $\frac{\partial}{\partial \mathbf{n}}$ indicates differentiation in the direction of the normal $\mathbf{n}$ of the boundary $\partial\Omega$ of the image domain Ω.

Let D be a unit-spacing grid over Ω and let the grid points be indexed by $\{1, 2, ..., N\}$ in lexicographical order. Let $a = fI_x$, $b = fI_y$, $c = -(xI_x + yI_y)$, $d = I_t$. For $\forall i \in \{1, 2, ..., N\}$, a discrete approximation of the Euler-Lagrange equations Eq. (6.61) is:

$$
\begin{aligned}
&a_i^2U_i + a_ib_iV_i + a_ic_iW_i + a_id_iZ_i + a_id_iZ_0 - \lambda\sum_{j\in\mathcal{N}_i}(U_j - U_i) = 0\\
&b_ia_iU_i + b_i^2V_i + b_ic_iW_i + b_id_iZ_i + b_id_iZ_0 - \lambda\sum_{j\in\mathcal{N}_i}(V_j - V_i) = 0\\
&c_ia_iU_i + c_ib_iV_i + c_i^2W_i + c_id_iZ_i + c_id_iZ_0 - \lambda\sum_{j\in\mathcal{N}_i}(W_j - W_i) = 0\\
&d_ia_iU_i + d_ib_iV_i + d_ic_iW_i + d_i^2Z_i + d_i^2Z_0 - \lambda\sum_{j\in\mathcal{N}_i}(Z_j - Z_i) = 0,
\end{aligned}
\tag{6.63}
$$

where $(U_i, V_i, W_i, Z_i) = (U, V, W, Z)_i$ is the scene flow vector at grid point i; a_i, b_i, c_i, d_i are the values at i of a, b, c, d, and $\mathcal{N}_i$ is the set of indices of the neighbors of i. The Laplacian ∇^2Q, $Q \in \{U, V, W, Z\}$ has been discretized using the 4-neighborhood as $\frac{1}{4}\sum_{j\in\mathcal{N}_i}(Q_j - Q_i)$, with the factor 1/4 absorbed by λ.

One can show that the matrix $\mathbf{A}$ of the resultant system of linear equations written for all grid points $i = 1, ..., N$, in this order, is positive definite [67], which means that the point-wise and block-wise Gauss-Seidel iterations converge [62, 63]. This is akin to optical flow estimation by the Horn and Schunck algorithm [68]. For a 4×4 block division of matrix $\mathbf{A}$, the Gauss-Seidel iterations consist of solving, for each $i \in \{1, ..., N\}$, the following 4×4 linear system of equations, where k is the iteration number:

$$
\begin{aligned}
(a_i^2 + \lambda n_i)U_i^{k+1} + a_i b_i V_i^{k+1} + a_i c_i W_i^{k+1} + a_i d_i Z_i^{k+1} &= r_U^{k+1} \\
b_i a_i U_i^{k+1} + (b_i^2 + \lambda n_i)V_i^{k+1} + b_i c_i W_i^{k+1} + b_i d_i Z_i^{k+1} &= r_V^{k+1} \\
c_i a_i U_i^{k+1} + c_i b_i V_i^{k+1} + (c_i^2 + \lambda n_i)W_i^{k+1} + c_i d_i Z_i^{k+1} &= r_W^{k+1} \\
d_i a_i U_i^{k+1} + d_i b_i V_i^{k+1} + d_i c_i W_i^{k+1} + (d_i^2 + \lambda n_i)Z_i^{k+1} &= r_Z^{k+1},
\end{aligned}
\tag{6.64}
$$

where the righthand side is defined by:

$$
\begin{aligned}
r_U^{k+1} &= -a_i d_i Z_0 + \lambda \left(\sum_{j \in \mathcal{N}_i; j<i} U_j^{k+1} + \sum_{j \in \mathcal{N}_i; j>i} U_j^k \right) \\
r_V^{k+1} &= -b_i d_i Z_0 + \lambda \left(\sum_{j \in \mathcal{N}_i; j<i} V_j^{k+1} + \sum_{j \in \mathcal{N}_i; j>i} V_j^k \right) \\
r_W^{k+1} &= -c_i d_i Z_0 + \lambda \left(\sum_{j \in \mathcal{N}_i; j<i} W_j^{k+1} + \sum_{j \in \mathcal{N}_i; j>i} W_j^k \right) \\
r_Z^{k+1} &= -d_i^2 Z_0 + \lambda \left(\sum_{j \in \mathcal{N}_i; j<i} Z_j^{k+1} + \sum_{j \in \mathcal{N}_i; j>i} Z_j^k \right).
\end{aligned}
\tag{6.65}
$$

The resolution of this 4×4 system can be done efficiently by the singular value decomposition method [50].

Example: This example uses the *Unmarked rocks* sequence from the CMU VASC image database (http://vasc.ri.cmu.edu//idb/html/motion/). The scene is static and

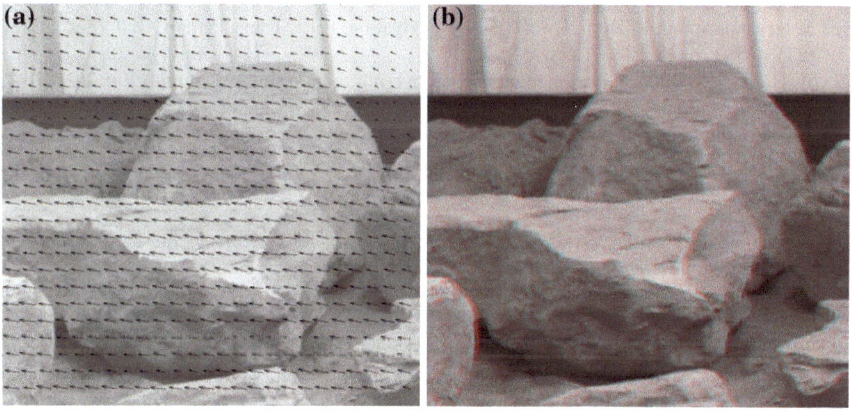

Fig. 6.9 *Unmarked rocks* sequence: **a** Optical flow reconstructed from the computed scene flow and depth; **b** an anaglyph of the *Unmarked rocks* scene constructed from the computed depth

the camera slides horizontally to the right, which means that the actual optical velocities over the image domain are approximately horizontal, directed to the left, and of about constant magnitude. There is no ground truth but we can indirectly assess the results via optical flow reconstructed from the scheme's scene flow and depth output. In this case, optical flow is computed using Eq. (6.3). The results are shown in Fig. 6.9a. A visual inspection and a comparison to the Horn and Schunck algorithm output in [67] confirms that the scene flow estimation scheme is valid and reliable. The results can also be judged by viewing the anaglyph of Fig. 6.9b. This anaglyph, to be viewed with red/blue glasses, gives a good impression of the scene structure.

References

1. J.J. Gibson, *The perception of the visual world* (Houghton Mifflin, Boston, 1950)
2. J.J. Gibson, Optical motions and transformations as stimuli for visual perception. Psychol. Rev. **64**, 288–295 (1957)
3. K. Prazdny, On the information in optical flows. Comput. Vis. Graph. Image Process. **22**, 239–259 (1983)
4. J.K. Aggarwal, N. Nandhakumar, On the computation of motion from a sequence of images: a review. Proc. IEEE **76**, 917–935 (1988)
5. T. Huang, A. Netravali, Motion and structure from feature correspondences: a review. Proc. IEEE **82**, 252–268 (1994)
6. A. Mitiche, *Computational Analysis of Visual Motion* (Plenum Press, New York, 1994)
7. R. Chellapa, A.A. Sawchuk, *Digital Image Processing and Analysis: Volume 2: Digital Image Analysis* (IEEE Computer Society Press, New York, 1985)
8. J.W. Roach, J.K. Aggarwal, Determining the movement of objects from a sequence of images. IEEE Trans. Pattern Anal. Mach. Intell. **2**(6), 554–562 (1980)
9. R.Y. Tsai, T.S. Huang, Uniqueness and estimation of three-dimensional motion parameters of rigid objects with curved surfaces. IEEE Trans. Pattern Anal. Mach. Intell. **6**(1), 13–27 (1984)
10. J. Aloimonos, C. Brown, Direct processing of curvilinear sensor motion from a sequence of perspective images, in *IEEE Workshop on Computer Vision: Representation and Analysis*, Annapolis, MD, 1984, pp. 72–77
11. S. Negahdaripour, B. Horn, Direct passive navigation. IEEE Trans. Pattern Anal. Mach. Intell. **9**(1), 168–176 (1987)
12. B. Horn, E. Weldon, Direct methods for recovering motion. Int. J. Comput. Vis. **2**(2), 51–76 (1988)
13. H.C. Longuet-Higgins, K. Prazdny, The interpretation of a moving retinal image. Proc. R. Soc. Lond. B **208**, 385–397 (1980)
14. H.C. Longuet-Higgins, A computer algorithm for reconstructing a scene from two projections. Nature **293**, 133–135 (1981)
15. X. Zhuang, R. Haralick, Rigid body motion and the optical flow image. in *First International Conference on Artificial Intelligence Applications*, 1984, pp. 366–375
16. S. Ullman, The interpretation of structure from motion. Proc. R. Soc. Lond. B **203**, 405–426 (1979)
17. B. Horn, B. Schunck, Determining optical flow. Artif. Intell. **17**, 185–203 (1981)
18. S. Solimini, J.M. Morel, *Variational Methods in Image Segmentation* (Springer, Boston, 2003)
19. S. Osher, N. Paragios (eds.), *Geometric Level Set Methods in Imaging, Vision, and Graphics* (Springer, New York, 2003)
20. G. Aubert, P. Kornpbrost, *Mathematical Problems in Image Processing: Partial Differential Equations and the Calculus of Variations* (Springer, New York, 2006)

21. A. Mitiche, I. Ben Ayed, *Variational and Level Set Methods in Image Segmentation* (Springer, New York, 2010)
22. A. Bruss, B. Horn, Passive navigation. Comput. Graph. Image Process. **21**, 3–20 (1983)
23. G. Adiv, Determining three-dimensional motion and structure from optical flow generated by several moving objects. IEEE Trans. Pattern Anal. Mach. Intell. **7**(4), 384–401 (1985)
24. B. Shahraray, M. Brown, Robust depth estimation from optical flow, in *International Conference on Computer Vision*, 1988, pp. 641–650
25. D. Heeger, A. Jepson, Subspace methods for recovering rigid motion I: algorithm and implementation. Int. J. Comput. Vis. **7**(2), 95–117 (1992)
26. M. Taalebinezhaad, Direct recovery of motion and shape in the general case by fixation. IEEE Trans. Pattern Anal. Mach. Intell. **14**(8), 847–853 (1992)
27. E. De Micheli, F. Giachero, Motion and structure from one dimensional optical flow, in *IEEE International Conference on Computer Vision and, Pattern Recognition*, 1994, pp. 962–965
28. N. Gupta, N. Kanal, 3-D motion estimation from motion field. Artif. Intell. **78**, 45–86 (1995)
29. Y. Xiong, S. Shafer, Dense structure from a dense optical flow, in *International Conference on Computer Vision and Image Understanding*, 1998, pp. 222–245
30. Y. Hung, H. Ho, A Kalman filter approach to direct depth estimation incorporating surface structure. IEEE Trans. Pattern Anal. Mach. Intell. **21**(6), 570–575 (1999)
31. S. Srinivasan, Extracting structure from optical flow using the fast error search technique. Int. J. Comput. Vis. **37**(3), 203–230 (2000)
32. T. Brodsky, C. Fermuller, Y. Aloimonos, Structure from motion: beyond the epipolar constraint. Int. J. Comput. Vis. **37**(3), 231–258 (2000)
33. H. Liu, R. Chellapa, A. Rosenfeld, A hierarchical approach for obtaining structure from two-frame optical flow, in *IEEE Workshop on Motion and Video, Computing*, 2002
34. W. MacLean, A. Jepson, R. Frecher, Recovery of egomotion and segmentation of independent object motion using the em algorithm. British Mach. Vis. Conf. BMVC **94**, 13–16 (1994)
35. S. Fejes,L. Davis, What can projections of flow fields tell us about visual motion, in *International Conference on Computer Vision*, 1998, pp. 979–986
36. J. Weber, J. Malik, Rigid body segmentation and shape description from dense optical flow under weak perspective. IEEE Trans. Pattern Anal. Mach. Intell. **19**(2), 139–143 (1997)
37. A. Mitiche, S. Hadjres, MDL estimation of a dense map of relative depth and 3D motion from a temporal sequence of images. Pattern Anal. Appl. **6**, 78–87 (2003)
38. H. Sekkati, A. Mitiche, A variational method for the recovery of dense 3D structure from motion. Robotics and Autonomous Systems **55**(7), 597–607 (2007)
39. H. Sekkati, A. Mitiche, Concurrent 3D-motion segmentation and 3D interpretation of temporal sequences of monocular images. IEEE Trans. Image Process. **15**(3), 641–653 (2006)
40. A. Mitiche, H. Sekkati, Optical flow 3D segmentation and interpretation: a variational method with active curve evolution and level sets. IEEE Trans. Pattern Anal. Mach. Intell. **28**(11), 1818–1829 (2006)
41. H. Sekkati, A. Mitiche, Joint optical flow estimation, segmentation, and 3D interpretation with level sets. Comput. Vis. Image Underst. **103**(2), 89–100 (2006)
42. Y.G. Leclerc, Constructing simple stable descriptions for image partitioning. Int. J. Comput. Vis. **3**(1), 73–102 (1989)
43. S. Vedula, S. Baker, P. Rander, R. Collins, T. Kanade, Three-dimensional scene flow. IEEE Trans. Pattern Anal. Mach. Intell. **27**, 475–480 (2005)
44. J.-P. Pons, R. Keriven, O. Faugeras, G. Hermosillo, Variational stereovision and 3D scene flow estimation with statistical similarity measures, in *International Conference on Computer Vision*, Nice, France, 2003, pp. 597–602
45. Y. Zhang, C. Kambhamettu, Integrated 3D scene flow and structure recovery from multiview image sequences, in *IEEE Conference on Computer Vision and Pattern Recognition*, vol. 2, pp. 674–681 (2000)
46. F. Huguet, F. Devernay, A variational method for scene flow estimation from stereo sequences, in *Proceedings of the International Conference on Computer Vision*, Rio de Janeiro, Brasil: IEEE, Oct 2007

47. A. Wedel, C. Rabe, T. Vaudrey, T. Brox, U. Franke, D. Cremers, Efficient dense scene flow from sparse or dense stereo data, in *European Conference on Computer Vision (ECCV)*, Marseille, France, Oct. 2008
48. F. Raudies, H. Neumann, A review and evaluation of methods estimating ego-motion. Comput. Vis. Image Underst. **116**(5), 606–633 (2012)
49. F. Raudies, H. Neumann, An efficient linear method for the estimation of ego-motion from optical flow, in *DAGM-Symposium*, 2009, pp. 11–20
50. G.E. Forsyth, A.A. Malcolm, C.B. Moler, *Computer Methods for Mathematical Computations* (Prentice Hall, Englewood Cliffs, 1977)
51. S. Negahdaripour, B. Horn, *Direct Passive Navigation* (Massachusetts Institute of Technology Memo 821, Cambridge, 1985)
52. S. Negahdaripour, A. Yuille, *Direct Passive Navigation: Analytical Solution for Quadratic Patches* (Massachusetts Institute of Technology Memo 876, Cambridge, 1986)
53. G. Aubert, G. Deriche, P. Kornprobst, Computing optical flow via variational thechniques. SIAM J. Appl. Math. **60**(1), 156–182 (1999)
54. H. Sekkati, A. Mitiche, Dense 3D interpretation of image sequences: A variational approach using anisotropic diffusion, in *International Conference on Image Analysis and Processing*, Mantova, Italy, 2003, pp. 424–429
55. T. Nir, A.M. Bruckstein, R. Kimmel, Over-parameterized variational optical flow. Int. J. Comput. Vis. **76**(2), 205–216 (2008)
56. R. Deriche, P. Kornprobst, G. Aubert, Optical-flow estimation while preserving its discontinuities: a variational approach, in *Asian Conference on Computer Vision*, 1995, pp. 71–80
57. J. Oliensis, A critique of structure-from-motion algorithms. Comput. Vis. Image Underst. **80**(2), 172–214 (2000)
58. S. Bougnoux, From projective to euclidean space under any practical situation, a criticism of self-calibration, in *ICCV*, 1998, pp. 790–798
59. E. Dubois, A projection method to generate anaglyph stereo images, in *International Conference on Acoustics, Speech, and Signal Processing*, 2001 vol. III, pp. 1661–1664
60. S. Osher, N. Paragios, *Geometric Level Set Methods in Imaging, Vision, and Graphics* (Birkhauser, Berlin, 1995)
61. R. Feghali, A. Mitiche, Fast computation of a boundary preserving estimate of optical flow. SME Vis. Q. **17**(3), 1–4 (2001)
62. J. Stoer, P. Burlisch, *Introduction to Numerical Methods*, 2nd ed. (Springer, New York, 1993)
63. P. Ciarlet, *Introduction à l'analyse numérique matricielle et à l'optimisation* (Masson, Fifth, 1994)
64. C. Debrunner, N. Ahuja, Segmentation and factorization-based motion and structure estimation for long image sequences. IEEE Trans. Pattern Anal. Mach. Intell. **20**(2), 206–211 (1998)
65. A. Mitiche, On combining stereopsis and kineopsis for space perception, in *IEEE Conference on Artificial Intelligence Applications*, 1984, pp. 156–160
66. A. Mitiche, A computational approach to the fusion of stereopsis and kineopsis, in *Motion Understanding: Robot and Human Vision*, ed. by W.N. Martin, J.K. Aggarwal (Kluwer Academic Publishers, Norwell1988), pp. 81–99
67. A. Mitiche, Y. Mathlouthi, I. Ben Ayed, A linear method for scene flow estimation from a single image sequence, in *INRS-EMT Technical report*, 2011
68. A. Mitiche, A. Mansouri, On convergence of the Horn and Schunck optical flow estimation method. IEEE Trans. Image Process. **13**(6), 848–852 (2004)

Index

A. Mitiche and J. K. Aggarwal, *Computer Vision Analysis of Image Motion by Variational Methods*, Springer Topics in Signal Processing 10, DOI: 10.1007/978-3-319-00711-3,